PREFACE

The first British Railways diesel shunter sold into industrial service was Barclay-built D2953, which went to Thames Matex Ltd of West Thurrock in June 1966. Many other organisations were quick to realise that BR's obsolete small shunters had several years useful life left in them, and could be obtained at relatively cheap prices when compared to buying new. In the six years following D2953's sale, no less than 188 diesel locomotives followed into industrial service, and these were recorded in the first edition of this book, published back in May 1972 by *The Industrial Railway Society.* After a further nine years the second edition detailed 282 locos sold out of service. This expanded third edition lists no less than 315 ex-BR locomotives. The information has been compiled utilising notes and observations provided by members of *The Industrial Railway Society,* and the book includes all information reported up to 31st May 1987. For the sake of completeness the small number of preserved shunters, of the same classes as sold to industry, are included. These are akin to their industrial cousins, but preserved passenger locos are not included as they are considered to be outside the scope of this book.

Over the years the search for rare ex-main line motive power has taken many BR enthusiasts to industrial sites. The additional sight of purpose built, purely industrial locos, at these locations has often resulted in main line enthusiasts developing an interest in this form of motive power. Such individuals may like to know, therefore, that *The Industrial Railway Society* caters for those people interested in private industrial railways and motive power. Benefits of Society membership include an illustrated printed magazine; site visits; occasional railtours; a well stocked library; and bi-monthly news bulletins giving all the latest information, together with separate periodical updates to this book. The author will be pleased to provide further details to prospective members.

Several IRS officers and members have assisted during the compilation of this volume, and the author wishes to thank Alan Bridges, Brian Cuttell, Bob Darvill, Roy Etherington, Robert Pritchard, Ken Scanes, David Strickland, Eric Tonks, and John Wade for their help, plus many members who have sent in observations. Thanks are also due to: the staff of several NCB Plant Records Offices in the South Yorkshire Area; the Engineer's Department at Manton Colliery; Peter Clayton at NCB Manvers; other colliery officials and men; the staff at various private companies; the photographers who have helped to provide a fairly comprehensive set of pictures; my wife Janet for her usual assistance with proofreading; and to Mrs K. Bates who typed the manuscript.

This updated edition was written in May 1987 at Rotherham, and at Arisaig on the beautiful West Coast of Scotland.

439 Herringthorpe Valley Road
Rotherham
South Yorkshire
S65 3AF

A.J. BOOTH
1st June 1987

EXPLANATORY NOTES

The columns of information will be simple to understand, but the following notes are pertinent:—

BR number

The 1957 numbering system (with the 'D' prefix) has been used to dictate the order of the classes in the book, although the later TOPS Computer numbers (where allocated) are also given.

Builders

Many locos were built at BR's own workshops at Ashford, Darlington, Derby, Doncaster, Horwich, and Swindon, and these are shown where appropriate. Locos which were constructed at private companies' workshops are indicated by standard IRS abbreviations, as under. In the case of Drewry locos two numbers are given: this company was basically a sales organisation who did not erect locomotives, but rather sub-contracted the construction work, in practice to RSH and VF. Both organisations' works numbers are given in these cases.

AB	Andrew Barclay, Sons & Co Ltd, Kilmarnock
CE	Clayton Equipment Co Ltd, Hatton, Derby
DC	Drewry Car Co Ltd, London
EE	English Electric Co Ltd, Newton le Willows
HC	Hudswell, Clarke & Co Ltd, Leeds
HE	Hunslet Engine Co Ltd, Leeds
NB	North British Locomotive Co Ltd, Glasgow
RH	Ruston & Hornsby Ltd, Lincoln
RSH	Robert Stephenson & Hawthorns Ltd, Newcastle upon Tyne
VF	Vulcan Foundry Ltd, Newton le Willows
YE	Yorkshire Engine Co Ltd, Sheffield

Works number

This only applies to locomotives built by private companies. Those constructed at BR workshops were not allocated individual works numbers.

Year built

The year quoted is that in which the locomotive was officially added to BR stock, or the works-plate date if this is known to be different.

Last BR shed

The old style BR shed code is given, as per the accompanying list, and these codes have been adopted for two reasons. Firstly, certain sheds listed did not survive to be allocated the new type of BR two-letter codes; and secondly they are used for all sheds for the sake of uniformity. Initials have been used to indicate the workshops, etc. Codes used are as follows:—

1E	Bletchley	8J	Allerton
2F	Bescot	9A	Longsight
5A	Crewe Diesel	9D	Newton Heath
6A	Chester West	10D	Lostock Hall
6G	Llandudno Junction	12A	Carlisle Diesel
8C	Speke Junction	12C	Barrow
8F	Springs Branch	15A	Leicester Midland
8H	Birkenhead	16A	Toton

BR in INDUSTRY

Full details of all British Railways Diesel Locomotives sold for Industrial Service, and Preservation – Past and present

A.J. BOOTH

HANDBOOK 3BRD

Third Edition, published in 1987 by the Industrial Railway Society, at 439 Herringthorpe Valley Road, Rotherham, South Yorkshire, S65 3AF.

ISBN 0 901096 56 3

Cover photograph: Following its transfer from Brookhouse Colliery, D2229 became a popular loco at Manton Colliery. It was photographed on 21st May 1987 pulling a rake of MGR wagons under the screens, with driver David Hall at the controls. (A.J. Booth)

16B	Colwick	64H	Leith Central
16C	Derby	65A	Eastfield
16F	Burton on Trent	66A	Polmadie
30A	Stratford	67C	Ayr
30E	Colchester	70D	Eastleigh
31A	Cambridge	70F	Bournemouth West
31B	March	70H	Ryde
32A	Norwich Thorpe	73C	Hither Green
34E	New England	73F	Chart Leacon
36A	Doncaster	75C	Selhurst
36C	Frodingham	81A	Old Oak Common
40A	Lincoln	81F	Oxford
40B	Immingham	82A	Bristol (Bath Road)
41A	Tinsley	82C	Swindon
41J	Shirebrook	83B	Taunton
50B	Hull Dairycoates	84A	Plymouth Laira
50C	Hull Botanic Gardens	85A	Worcester
50D	Goole	85B	Gloucester
51A	Darlington	86A	Cardiff Canton
51L	Thornaby	87E	Swansea Landore
52A	Gateshead	BSD	Beeston Sleeper Depot
55B	York	CJ	Chesterton Junction
55C	Healey Mills	CW	Crewe Works
55F	Bradford (Hammerton Street)	HW	Horwich Works
55G	Knottingley	RSD	Reading Signal Depot
62A	Thornton	SW	Swindon Works
62C	Dunfermline	WW	Wolverton Works

Date withdrawn

The official date withdrawn from BR stock.

P/F and Title

These columns indicate whether the loco title is the Present (P) or Former (F) one carried by the locomotive. NPT indicates 'No present title', meaning the loco presently carries neither name nor number. Former titles given for scrapped locomotives are those carried by the loco at the time it was cut up. NFT indicates 'No former title' meaning the loco carried neither name nor number at the time it was scrapped.

Notes

These follow on beneath the basic loco data. The first industrial user is stated, together with the date the loco arrived from BR. In certain cases the move was protracted, for example due to a 'hot box', and these are indicated by a period, such as 'June to September 1969'. This means the loco left its BR Depot in June 1969, but did not arrive at its destination until September 1969. Locos were moved from their last BR shed unless otherwise stated. All known subsequent movements and disposals are given, with the dates where possible. Certain abbreviations are used:—

APCM	Associated Portland Cement Manufacturers Ltd
BR	British Railways
BSC	British Steel Corporation
CEGB	Central Electricity Generating Board
NCB	National Coal Board
NCBOE	National Coal Board Opencast Executive
NSF	National Smokeless Fuels Ltd

Companies frequently change their titles and ownership. These are not recorded, however, if the loco remained at the same site.

Layout

The book is divided into 24 sections, each devoted to one class, and basic details appropriate to each section are given below. The dimensions, etc, apply to locomotives as built, and not necessarily as when sold by BR. The sections are followed by three appendices.

SECTION 1

British Railways built 0-6-0 diesel mechanical locomotives, numbered D2000–D2199, and D2372–D2399, introduced 1957. Fitted with a Gardner type 8L3 engine developing 204bhp at 1200rpm; maximum tractive effort of 15,650lb; five speed gearbox; weight 30tons 16cwt; driving wheel diameter 3ft 7in. Later reclassified TOPS Class 03.

SECTION 2

Drewry Car Co Ltd 0-6-0 diesel mechanical locomotives built by Vulcan Foundry Ltd, numbered D2200–D2214, introduced 1952. Fitted with a Gardner type 8L3 engine developing 204bhp at 1200rpm; maximum tractive effort of 16,850lb; five speed gearbox; weight 29tons 15cwt; driving wheel diameter 3ft 3in; Later reclassified TOPS Class 04.

SECTION 3

Drewry Car Co Ltd 0-6-0 diesel mechanical locomotives built by Vulcan Foundry Ltd and Robert Stephenson & Hawthorns Ltd, numbered D2215–D2273, introduced 1955. Fitted with a Gardner type 8L3 engine developing 204bhp at 1200rpm; maximum tractive effort 15,650lb; five speed gearbox; weight 29tons 15cwt; driving wheel diameter 3ft 6in. Later reclassified TOPS Class 04.

SECTION 4

Drewry Car Co Ltd 0-6-0 diesel mechanical locomotives built by Robert Stephenson & Hawthorns Ltd, numbered D2274–D2340, introduced 1959. Fitted with a Gardner type 8L3 engine developing 204bhp at 1200rpm; maximum tractive effort 16,850lb; five speed gearbox; weight 29tons 15cwt; driving wheel diameter 3ft 7in. Later reclassified TOPS Class 04. Departmental DS1173, built in 1947, was later renumbered D2341 and completed Class 04, but some particulars were different from those of the main batch.

SECTION 5

Andrew Barclay, Sons & Co Ltd built 0-4-0 diesel mechanical locomotives, numbered D2410–D2444, introduced 1958. Fitted with a Gardner type 8L3 engine developing 204bhp at 1200rpm; maximum tractive effort 20,000lb; five speed gearbox; weight 35tons; driving wheel diameter 3ft 7in. Later reclassified TOPS Class 06.

SECTION 6

Hudswell, Clarke & Co Ltd built 0-6-0 diesel mechanical locomotives, numbered D2510–D2519, introduced 1961. Fitted with a Gardner type 8L3 engine developing 204bhp at 1200rpm; maximum tractive effort 16,100lb; four speed gearbox; weight 34tons 4cwt; driving wheel diameter 3ft 6in. No TOPS classification.

SECTION 7

Hunslet Engine Co Ltd built 0-6-0 diesel mechanical locomotives, numbered D2550–D2618, introduced 1955. Fitted with a Gardner type 8L3 engine developing 204bhp at 1200rpm; maximum tractive effort 14,500lb; four speed gearbox; weight 30tons; driving wheel diameter 3ft 4in (D2550–D2573), and 3ft 9in (D2574–D2618). Later reclassified TOPS Class 05.

SECTION 8

North British Locomotive Co Ltd built 0-4-0 diesel hydraulic locomotives, numbered D2708–

D2780, introduced 1957. Fitted with a North British/M.A.N. type W6V 17.5/22A engine developing 225bhp at 1100rpm; maximum tractive effort 20,080lb; weight 30tons; driving wheel diameter 3ft 6in. No TOPS classification.

SECTION 9

Yorkshire Engine Co Ltd built 0-4-0 diesel hydraulic locomotives, numbered D2850–D2869, introduced 1960. Fitted with a Rolls-Royce type C6NFL engine developing 179bhp at 1800rpm; maximum tractive effort 15,000lb; weight 28tons; driving wheel diameter 3ft 6in. Later reclassified TOPS Class 02.

SECTION 10

Hunslet Engine Co Ltd built 0-4-0 diesel mechanical locomotives, numbered D2950–D2952, introduced 1954. Fitted with a Gardner type 6L3 engine developing 153bhp at 1200rpm; maximum tractive effort 10,800lb; four speed gearbox; weight 22tons 9cwt; driving wheel diameter 3ft 4in. No TOPS classification.

SECTION 11

Andrew Barclay, Sons & Co Ltd built 0-4-0 diesel mechanical locomotives, numbered D2953–D2956, introduced 1956. Fitted with a Gardner type 6L3 engine developing 153bhp at 1200rpm; maximum tractive effort 12,750lb; four speed gearbox; weight 25tons; driving wheel diameter 3ft 2in. BR Departmental locomotive number 81 became the second D2956 after the first was withdrawn. Later reclassified TOPS Class 01.

SECTION 12

Ruston & Hornsby Ltd built 0-4-0 diesel mechanical locomotives, numbered D2957–D2958, introduced 1956. Fitted with a Ruston type 6VPHL engine developing 165bhp at 1250rpm; maximum tractive effort 14,350lb; weight 28tons; driving wheel diameter 3ft 4in. No TOPS classification.

SECTION 13

Ruston & Hornsby Ltd built 0-6-0 diesel electric locomotives, numbered D2985–D2998, introduced 1962. Fitted with a Paxman type 6RPHL engine developing 275bhp at 1360rpm; maximum tractive effort 28,240lb; weight 42tons 5cwt; driving wheel diameter 3ft 6in. Later reclassified TOPS Class 07.

SECTION 14

British Railways built 0-6-0 diesel electric locomotives, numbered D3000–D3116, introduced 1953. Fitted with an English Electric type 6KT engine developing 350bhp at 630rpm; maximum tractive effort 35,000lb; weight 49tons; driving wheel diameter 4ft 6in. Later reclassified TOPS Class 08.

SECTION 15

British Railways built 0-6-0 diesel electric locomotives, numbered D3127–D3136, D3167–D3438, D3454–D3472, D3503–D3611, D3652–D4048, and D4095–D4192, introduced 1953. Fitted with an English Electric type 6KT engine developing 350bhp at 680rpm; maximum tractive effort 35,000lb; weight 48tons; driving wheel diameter 4ft 6in. Later reclassified TOPS Class 08.

SECTION 16

British Railways built 0-6-0 diesel electric locomotives, numbered D3137–D3151, D3439–D3453; D3473–D3502; D3612–D3651; and D4049–D4094, introduced 1955. Fitted with a Blackstone type ER6T engine developing 350bhp at 750rpm; maximum tractive effort 35,000lb; weight 47tons 10cwt; driving wheel diameter 4ft 6in. Later reclassified TOPS Class 10.

SECTION 17

Clayton Equipment Co Ltd built Bo-Bo diesel electric locomotives, numbered D8500–D8616, introduced 1962. Fitted with two Paxman type 6ZHXL engines, each developing 450bhp at 1500rpm; maximum tractive effort 40,000lb; weight 68tons; driving wheel diameter 3ft 3½in. Later reclassified TOPS Class 17, with three sub-divisions of which D8568 was 17/1.

SECTION 18

British Railways built 0-6-0 diesel hydraulic locomotives, numbered D9500–D9555, introduced 1964. Fitted with a Paxman 'Ventura' type 6YJX engine developing 650bhp at 1500rpm; maximum tractive effort 30,910lb; weight 50tons; driving wheel diameter 4ft 0in. Later reclassified TOPS Class 14.

SECTION 19

LMS and British Railways built 0-6-0 diesel electric locomotives, numbered 12033–12138, introduced 1945. Fitted with an English Electric type 6KT engine developing 350bhp at 680rpm; maximum tractive effort 35,000lb; weight 47tons 5cwt; driving wheel diameter 4ft 0½in. Later reclassified TOPS Class 11.

SECTION 20

British Railways built 0-6-0 diesel electric locomotives, numbered 15211–15236, introduced 1949. Fitted with an English Electric type 6KT engine developing 350bhp at 680rpm; maximum tractive effort 24,000lb; weight 45tons; driving wheel diameter 4ft 6in. Later reclassified TOPS Class 12.

SECTION 21

Ruston & Hornsby Ltd built, 3ft 0in gauge, 4-wheel diesel mechanical locomotive, number ED10, built 1958 to the maker's Class 48DS. Fitted with a Ruston type 4YC engine developing 48bhp at 1375rpm; maximum tractive effort 3920lb; three speed gearbox; weight 7½tons; driving wheel diameter 2ft 6in. No TOPS classification.

SECTION 22

Ruston & Hornsby Ltd built, 1ft 6in gauge, 4-wheel diesel mechanical locomotive, number ZM32, built 1957 to the maker's Class LAT. Fitted with a Ruston type 2VSH engine developing 20bhp at 1200rpm; maximum tractive effort 1890lb; two speed gearbox; weight 3½tons; driving wheel diameter 1ft 4.5/16in. No TOPS classification.

SECTION 23

English Electric Ltd built 0-6-0 diesel electric locomotive, number D0226, built 1957. Fitted with an English Electric type 6RKT engine developing 500bhp at 750rpm; maximum tractive effort 33,000lb; weight 48tons; driving wheel diameter 4ft 0in. This loco was tested by BR but was never incorporated into capital stock. No TOPS classification.

SECTION 24

Ruston & Hornsby Ltd built 4-wheel diesel mechanical locomotives, of the maker's LB class. Fitted with Ruston type 3VSH engines of 31bhp at 1800rpm, and 1ft 4¼in diameter wheels. Ruston's built no less than 557 examples of this class, to 23 different narrow gauges, with the two examples recorded in this book being of 2ft 0in gauge. No TOPS classification.

SUMMARY

The total number of ex-BR locomotives sold to industry and preservation in the UK is as follows:—

Section		
Section	1	53
	2	8
	3	21
	4	32
	5	2
	6	4
	7	16
	8	10
	9	11
	10	1
	11	3
	12	1
	13	9
	14	19
	15	28
	16	17
	17	1
	18	48
	19	23
	20	3
	21	1
	22	1
	23	1
	24	2
	Total	315

BR number	Builder	Works number	Year built	Last BR Shed	Date wdn	P/F	Title
SECTION 1							
D2012	Swindon		1958	31B	12/75	P	03012/F135L
03012	to A. King & Sons Ltd, Snailwell, Cambridgeshire, 28th July 1976.						
D2018	Swindon		1958	32A	11/75	P	No.2/600
03018	to George Cohen, Sons & Co Ltd, Cransley, Northamptonshire, April 1976; to 600 Fragmentisers Ltd, Willesden, London, October 1980.						
D2020	Swindon		1958	32A	12/75	P	03020/F134L
03020	to A. King & Sons Ltd, Snailwell, Cambridgeshire, July 1976.						
D2022	Swindon		1958	52A	11/82	P	03022
03022	to Swindon & Cricklade Railway, 18th November 1983.						
D2023	Swindon		1958	40A	7/71	P	3
	to Tees & Hartlepool Port Authority, Middlesbrough Docks, July 1972; to Grangetown Docks, 15th September 1980; to Kent & East Sussex Railway, 14th August 1983.						
D2024	Swindon		1958	40A	7/71	P	4
	to Tees & Hartlepool Port Authority, Middlesbrough Docks, July 1972; to Grangetown Docks, 15th September 1980; to Kent & East Sussex Railway, 4th September 1983.						
D2027	Swindon		1958	30E	1/76	P	03027
03027	to Shipbreakers (Queenborough) Ltd, Kent, July 1976.						
D2037	Swindon		1959	32A	9/76	P	03037
03037	to Hargreaves Industrial Services Ltd, NCBOE British Oak Disposal Point, Crigglestone, October 1977; to West Hallam Disposal Point, November 1983; to British Oak, November 1984.						
D2041	Swindon		1959	75C	2/70	P	No.1
03041	to CEGB Richborough, February 1970; to CEGB Rye House, Hoddesdon, Hertfordshire, March 1971; to CEGB Barking about May 1971; to CEGB Rye House, August 1974; to Colne Valley Railway, Essex, 15th January 1981.						
D2046	Doncaster		1958	51L	10/71	P	2
	Rebuilt by Hunslet (6644 of 1967); to Gulf Oil Co Ltd, Waterston, Milford Haven, Pembrokeshire, May 1972.						
D2049	Doncaster		1958	50D	8/71	F	D2049
	to Hargreaves Industrial Services Ltd, NCBOE Bowers Row Disposal Point, Astley, West Yorkshire by 12th May 1974; to British Oak Disposal Point, Crigglestone, March 1975; to West Hallam Disposal Point, Mapperley, Derbyshire, March 1978; to British Oak, 28th January 1985; scrapped on site by Wath Skip Hire Ltd of Rotherham, November 1985.						
D2051	Doncaster		1959	30E	12/72	P	4
	to Ford Motor Co Ltd, Dagenham, London, about September 1973; rebuilt at BR Swindon Works from 3rd May 1977 to February 1978, and returned to Dagenham.						
D2054	Doncaster		1959	55B	11/72	F	CENTA
	to Chair Centre Ltd, Derby, October 1973; to British Industrial Sand Ltd, Middleton Towers, Norfolk, 13th July 1979; scrapped at C.F. Booth Ltd, Rotherham, September 1982.						
D2057	Doncaster		1959	51L	10/71	F	No.1
	Rebuilt by Hunslet (6645 of 1967); to NCB Grimethorpe Colliery, Barnsley, September 1972; to C.F. Booth Ltd, Rotherham, 25th March 1986; scrapped April 1986.						
D2062	Doncaster		1959	32A	12/80	P	03062
03062	to Dean Forest Railway Society Ltd, 30th September 1982.						
D2069	Doncaster		1959	52A	11/83	P	03069
03069	to Vic Berry Ltd, Leicester, 4th January 1984.						

BR number	Builder	Works number	Year built	Last BR Shed	Date wdn	P/F	Title
D2070	Doncaster		1959	51L	11/71	F	D2070. NPT
to Shipbreakers (Queenborough) Ltd, Kent, June 1972.							
D2072	Doncaster		1959	51A	3/81	P	03072
03072	to Lakeside & Haverthwaite Railway Co Ltd, Cumbria, August 1981.						
D2090	Doncaster		1960	55B	7/76	P	03090
03090	to National Railway Museum, York, August 1976.						
D2093	Doncaster		1960	51L	10/71	F	No.2
Rebuilt by Hunslet (6643 of 1967); to NCB Grimethorpe Colliery, Barnsley, September 1972; to C.F. Booth Ltd, Rotherham, 26th March 1986; scrapped April 1986.							
D2099	Doncaster		1960	51L	2/76	F	03099. NPT
03099	to National Smokeless Fuels, Fishburn Coking Plant, County Durham, July 1976; to Monkton Coking Plant, 26th February 1981.						
D2113	Doncaster		1960	55B	8/75	F	03113. NPT
03113	to Gulf Oil Co Ltd, Waterston, Milford Haven, 17th May 1976.						
D2114	Swindon		1959	82C	5/68	F	D2114
to Birds Commercial Motors Ltd, Long Marston, Worcestershire, August 1968; scrapped January 1975.							
D2117	Swindon		1959	8F	10/71	P	No.8
to Lakeside & Haverthwaite Railway Co Ltd, Cumbria, April 1972.							
D2118	Swindon		1959	12C	6/72	F	D2118. NPT
to Anglian Building Products Ltd, Lenwade, Norfolk, about August 1973.							
D2122	Swindon		1959	82A	11/72	F	D2122
to J. Cashmore Ltd, Newport, 1974; to Briton Ferry Steel Co Ltd, Glamorgan, 13th August 1974, in a dismantled state; used for spares only; remains scrapped by August 1976.							
D2123	Swindon		1959	82C	12/68	F	NFT
to Birds Commercial Motors Ltd, Long Marston, Worcestershire, June 1969; to Birds, Cardiff, April 1970; used as a stationary machine; to Birds, Bristol, 1971 for storage; later used as a stationary generator; still there in November 1978, in a dismantled state, but scrapped shortly afterwards.							
D2125	Swindon		1959	82C	12/68	F	NFT
to Birds (Swansea) Ltd, 40 Acre Site, Cardiff, May 1969; despatched from 87A; scrapped June 1976.							
D2132	Swindon		1960	82C	5/69	F	D2132/LESLEY
to NCB Bestwood Colliery, Nottinghamshire, October 1970; to New Hucknall Colliery, July 1971; to Pye Hill Colliery, 1981; to C.F. Booth Ltd, Rotherham, 27th November 1984; scrapped over next two days.							
D2133	Swindon		1960	82A	7/69	P	D2133
to British Cellophane Ltd, Bridgwater, Somerset, July 1969.							
D2138	Swindon		1960	82C	5/69	P	D2138
to NCB Bestwood Colliery, Nottinghamshire, October 1970; to Pye Hill Colliery, April 1971; to BR Toton for wheel turning, June 1978; returned to Pye Hill; to Midland Railway Company, Butterley, Derbyshire, 20th August 1985.							
D2139	Swindon		1960	85A	5/68	P	1
to A.R. Adams Ltd, Newport, December 1968; used as hire loco (see Appendix A); sold to NSF Coed Ely Coking Plant, Tonyrefail, Mid Glamorgan, by March 1971; to BREL Swindon Works for repairs, 31st March 1981 to 2nd March 1982; returned to Coed Ely; to Monkton Coking Plant, Hebburn, Tyne & Wear, December 1983.							
D2146	Swindon		1961	87A	9/68	F	D2146
to Birds Commercial Motors Ltd, Long Marston, Worcestershire, June 1969; scrapped July 1978.							

<table>
<tr><th>BR number</th><th>Builder</th><th>Works number</th><th>Year built</th><th>Last BR Shed</th><th>Date wdn</th><th>P/F</th><th>Title</th></tr>
<tr><td>D2148</td><td>Swindon</td><td></td><td>1960</td><td>55C</td><td>11/72</td><td>F</td><td>D2148. NPT.</td></tr>
<tr><td colspan="8">to Hargreaves Industrial Services Ltd, NCBOE Bowers Row Disposal Point, Astley, Yorkshire, August 1973; to Lindley Plant Ltd, NCBOE Gatewen Disposal Point, Denbighshire, September 1973; to Bowers Row, December 1973; following collision damage to original cab, 03149's cab purchased from Doncaster Works, and fitted on site; to Steamport Transport Museum, Southport, Merseyside, 14th March 1987.</td></tr>
<tr><td>D2150</td><td>Swindon</td><td></td><td>1960</td><td>55B</td><td>11/72</td><td>F</td><td>D2150. NPT.</td></tr>
<tr><td colspan="8">to British Salt Ltd, Middlewich, Cheshire, May 1973.</td></tr>
<tr><td>D2176</td><td>Swindon</td><td></td><td>1961</td><td>CW</td><td>5/68</td><td>F</td><td>D2176</td></tr>
<tr><td colspan="8">to George Cohen, Sons & Co Ltd, Cransley, Northamptonshire, October 1968; scrapped November 1971.</td></tr>
<tr><td>D2178</td><td>Swindon</td><td></td><td>1962</td><td>81F</td><td>9/69</td><td>P</td><td>No.2</td></tr>
<tr><td colspan="8">to A.R. Adams Ltd, Newport, Monmouthshire, February 1970; used as a hire loco (see Appendix A); sold to National Smokeless Fuels, Coed Ely Coking Plant, Tonyrefail, Mid Glamorgan, about May 1974; to BR Swindon Works, August 1979; returned to Coed Ely about February 1980; to Caerphilly Railway Preservation Society, 12th November 1985.</td></tr>
<tr><td>D2180
03180</td><td>Swindon</td><td></td><td>1962</td><td>32A</td><td>3/84</td><td>P</td><td>03180</td></tr>
<tr><td colspan="8">to Mayer Newman Ltd, Snailwell, Newmarket, Cambridgeshire, 26th July 1984.</td></tr>
<tr><td>D2181</td><td>Swindon</td><td></td><td>1962</td><td>87A</td><td>5/68</td><td>F</td><td>PRIDE OF GWENT</td></tr>
<tr><td colspan="8">to A.R. Adams Ltd, Newport, Monmouthshire, December 1968; despatched from 85A; used as a hire loco (see Appendix A); sold to Gwent Coal Distribution Centre, Newport, Monmouthshire, by August 1971; to Marple & Gillott Ltd, Sheffield, December 1986; scrapped January 1987.</td></tr>
<tr><td>D2182</td><td>Swindon</td><td></td><td>1962</td><td>87A</td><td>5/68</td><td>P</td><td>3/3</td></tr>
<tr><td colspan="8">to A.R. Adams Ltd, Newport, Monmouthshire, November to December 1968; despatched from 85A; used as a hire loco (see Appendix A); to Sir Lindsay Parkinson & Co Ltd, Glyn Neath, February 1969; to Lindley Plant Ltd, Gatewen Disposal Point, Denbighshire, September 1973; to NCBOE Bennerley Disposal Point, 1981; to NCBOE Wentworth Stores, Rotherham, 18th March 1982; to Bennerley, 6th May 1983; to Coalfield Farm Disposal Point, July 1983.</td></tr>
<tr><td>D2184</td><td>Swindon</td><td></td><td>1962</td><td>87A</td><td>12/68</td><td>P</td><td>D2184</td></tr>
<tr><td colspan="8">to Co-operative Wholesale Society Ltd, Coal Concentration Depot, Southend, August 1969; despatched from 85A; to Colne Valley Railway, Essex, 17th October 1986.</td></tr>
<tr><td>D2185</td><td>Swindon</td><td></td><td>1962</td><td>87A</td><td>12/68</td><td>F</td><td>D2185</td></tr>
<tr><td colspan="8">to Birds Commercial Motors Ltd, Long Marston, Worcestershire, May 1969; despatched from 85A; to Birds, Cardiff, 1972; to Birds, Long Marston, January 1978; scrapped June 1978.</td></tr>
<tr><td>D2186</td><td>Swindon</td><td></td><td>1962</td><td>81F</td><td>9/69</td><td>F</td><td>D2186</td></tr>
<tr><td colspan="8">to A.R. Adams Ltd, Newport, Monmouthshire, February 1970; used as a hire loco (see Appendix A); scrapped January 1981.</td></tr>
<tr><td>D2187</td><td>Swindon</td><td></td><td>1961</td><td>82C</td><td>5/68</td><td>F</td><td>D2187</td></tr>
<tr><td colspan="8">to Birds Commercial Motors Ltd, Long Marston, Worcestershire, September 1968; scrapped June 1978.</td></tr>
<tr><td>D2188</td><td>Swindon</td><td></td><td>1961</td><td>83B</td><td>5/68</td><td>F</td><td>D2188</td></tr>
<tr><td colspan="8">to Birds Commercial Motors Ltd, Long Marston, Worcestershire, September 1968; despatched from 82C; scrapped February 1978.</td></tr>
<tr><td>D2192</td><td>Swindon</td><td></td><td>1961</td><td>82C</td><td>1/69</td><td>P</td><td>No.2</td></tr>
<tr><td colspan="8">to Dart Valley Railway, Devon, August to September 1970; to Torbay & Dartmouth Railway, Paignton, July 1977.</td></tr>
</table>

BR number	Builder	Works number	Year built	Last BR Shed	Date wdn	P/F	Title
D2193	Swindon		1961	82C	1/69	F	2

to A.R. Adams Ltd, Newport, Monmouthshire, September to October 1969; despatched from 85A; used as a hire loco (see Appendix A); scrapped January 1981.

D2194	Swindon		1961	85A	9/68	F	D2194

to Birds Commercial Motors Ltd, Long Marston, Worcestershire, May 1969; scrapped July 1978.

D2195	Swindon		1961	82A	9/68	F	D10

to Llanelli Steel Co Ltd, Carmarthenshire, June 1969; despatched from 85A; sold per R.E. Trem Ltd, Finningley, Doncaster; scrapped about September 1981.

D2196	Swindon		1961	8H	6/83	P	No.40/GLYNIS JOYCE
03196	to R.O. Hodgson Ltd, Carnforth, 15th June 1983.						

D2199	Swindon		1961	12C	6/72	P	1

to BR Doncaster Works for overhaul; to NCB Rockingham Colliery, Birdwell, Barnsley, February 1974; to Barrow Colliery, Worsborough, Barnsley, about January 1979; to Houghton Main Colliery, Barnsley, about June 1979; to Royston Drift Mine, Barnsley, August 1980; to Barrow Colliery, 8th July 1981; to Royston Drift Mine, 23rd March 1982.

D2373	Swindon		1961	9D	5/68	F	No.1/DAWN

to NCB Manvers Coal Preparation Plant, Wath-on-Dearne, Rotherham, September 1968; despatched from 9K; to Barnburgh Colliery, July 1971; to Manvers, April 1976; to Wath Colliery, about March 1977; to Manvers, about September 1977; to Wath Colliery, about July 1978; to Manvers, November 1978; scrapped on site by E. Nortcliffe Ltd of Rotherham, 1982.

D2381	Swindon		1961	16C	6/72	P	D2381

to Flying Scotsman Enterprises, Market Overton, Rutland, April 1973; to Steamtown, Carnforth, 19th March 1976.

SECTION 2

D2203	DC VF	2400 D145	1952	CW	12/67	P	D2203

to Hemel Hempstead Lightweight Concrete Co Ltd, Cupid Green, Hertfordshire, February 1968; to Yorkshire Dales Railway, 8th February 1982.

D2204	DC VF	2485 D211	1953	55F	10/69	F	D5

to Briton Ferry Steel Co Ltd, Glamorgan, March 1970; sold via W & F Smith Ltd, Ecclesfield, Sheffield; scrapped September 1979.

D2205	DC VF	2486 D212	1953	51L	7/69	P	1

to Tees & Hartlepool Port Authority, Middlesbrough Docks, July 1970; to Kent & East Sussex Railway, 21st August 1983.

D2207	DC VF	2482 D208	1953	CW	12/67	P	D2207

to Hemel Hempstead Lightweight Concrete Co Ltd, Cupid Green, Hertfordshire, February 1968; to North Yorkshire Moors Railway, September 1973.

D2208	DC VF	2483 D209	1953	5A	7/68	F	D2208

to NCB Manvers Coal Preparation Plant, Wath-on-Dearne, Rotherham, November 1968; to Cortonwood Colliery, Wombwell, March 1969; to Cadeby Colliery, Conisbrough, by May 1969; to Silverwood Colliery about December 1970; dismantled June 1976; scrapped May 1979.

BR number	Builder	Works number	Year built	Last BR Shed	Date wdn	P/F	Title
D2209	DC VF	2484 D210	1953	8J	7/68	F	No.16/TRACEY

to NCB Manvers Coal Preparation Plant, Wath-on-Dearne, Rotherham, November 1968; to Kiveton Park Colliery, 1974; cannibalised for spares, and dismantled 1982; scrapped on site, Autumn 1985.

D2211	DC VF	2509 D243	1954	16C	7/70	F	WILF CLEMENT

to Powell Duffryn Fuels Ltd, NCBOE Coed Bach Disposal Point, August 1970; to Rees Industries Ltd, Llanelli, 3rd August 1978; scrapped about November 1980.

D2213	DC VF	2529 D257	1954	8H	8/68	F	D2213

to NCB Manvers Coal Preparation Plant, Wath-on-Dearne, Rotherham, September 1969; used for spares; remains scrapped February 1978.

SECTION 3

D2219	DC VF	2542 D268	1955	8H	4/68	F	D2219

to Barnsley District Coking Co Ltd, Barrow Coking Plant, Barnsley, October 1968; loaned to NCB Barrow Colliery in August 1969, and returned; scrapped May 1977.

D2225	DC VF	2548 D274	1955	8F	3/69	F	D2225/DEBRA

to NCB Manvers Coal Preparation Plant, Wath-on-Dearne, Rotherham, January 1970; to Wath Colliery, 8th December 1976; scrapped on site by Wath Skip Hire Ltd, July 1985.

D2228	DC VF	2551 D277	1955	8F	7/68	F	D2228/4

to Bowaters UK Paper Co Ltd, Sittingbourne, 17th February 1969; scrapped January to March 1979.

D2229	DC VF	2552 D278	1955	52A	12/69	P	D2229/No.5

to NCB Brookhouse Colliery, Beighton, 28th August 1970; despatched from 51L; to Orgreave Colliery 1971; to Brookhouse Colliery by April 1972; to Orgreave Colliery about October 1973; to Brookhouse Colliery by July 1974; to Manton Colliery, 28th March 1983.

D2238	DC VF	2562 D288	1955	8H	7/68	F	D2238/CAROL

to NCB Manvers Coal Preparation Plant, Wath-on-Dearne, Rotherham, November 1968; loaned for several months during 1975 to Coventry Home Fire Plant, Keresley; returned to Manvers; scrapped on site by E. Nortcliffe Ltd of Rotherham, 1982.

D2239	DC VF	2563 D289	1955	75C	9/71	F	NFT

to NCB Dodworth Colliery, Barnsley, September 1972; to C.F. Booth Ltd, Rotherham, 20th March 1986; scrapped March 1986.

D2241	DC VF	2565 D291	1956	30E	5/71	F	2241

to George Cohen, Sons & Co Ltd, Cransley, Northamptonshire, September 1971; scrapped November 1976.

D2243	DC RSH	2575 7862	1956	51L	7/69	F	MD2

to Tees & Hartlepool Port Authority, Middlesbrough Docks, July 1970; dismantled 1972; scrapped March 1973.

BR number	Builder	Works number	Year built	Last BR Shed	Date wdn	P/F	Title
D2244	DC RSH	2576 7863	1956	55F	6/70	F	5
to A.R. Adams Ltd, Newport, Monmouthshire, July 1970; used as a hire loco (see Appendix A); scrapped January 1981.							
D2245	DC RSH	2577 7864	1956	50D	12/68	P	No.2
to Derwent Valley Light Railway, Layerthorpe, York, May 1969; to Shackerstone Railway, Leicestershire, 17th May 1978.							
D2246	DC RSH	2578 7865	1956	55G	7/68	P	NPT
to Coal Mechanisation Ltd, Crawley, Sussex, January 1969.							
D2247	DC RSH	2579 7866	1956	55B	11/69	F	D6
to Briton Ferry Steel Co Ltd, Glamorgan, June 1970; sold via W & F Smith Ltd, Ecclesfield, Sheffield; scrapped September 1979.							
D2248	DC RSH	2580 7867	1957	55F	6/70	P	2243/No.18/SUE
to NCB Manvers Coal Preparation Plant, Wath-on-Dearne, Rotherham, June 1970; to Maltby Colliery about September 1971; following a repaint at Maltby the loco received the incorrect number 2243.							
D2258	DC RSH	2602 7879	1957	16C	9/70	F	D2258/4–2
to Hargreaves Industrial Services Ltd, NCBOE Shilo Disposal Point, Ilkeston, January 1971; to BR Toton for repairs, December 1974; returned to Shilo 1975; to NCBOE Wentworth Stores, Rotherham, 17th February 1984; to C.F. Booth Ltd, Rotherham, 2nd September 1986; scrapped January 1987.							
D2259	DC RSH	2603 7889	1957	73F	12/68	F	D2259/5
to Bowaters UK Paper Co Ltd, Sittingbourne, Kent, February 1969; scrapped January 1978.							
D2260	DC RSH	2604 7890	1957	55F	10/70	F	THOMAS HARLING
to Powell Duffryn Fuels Ltd, NCBOE Mill Pit Disposal Point, Cefn Cribbwr, July 1971; sold via Tilsley & Lovatt Ltd, Trentham; to Cwm Mawr Disposal Point, 3rd November 1981; to Coed Bach Disposal Point, December 1982; scrapped on site by Rees Industries Ltd of Llanelli, June 1983.							
D2262	DC RSH	2606 7892	1957	51A	9/68	F	7
to Ford Motor Co Ltd, Dagenham, London, March 1969; to BR Swindon Works for rebuild from 6th October 1977 to 24th February 1978; scrapped July 1978.							
D2267	DC RSH	2611 7897	1957	50D	12/69	P	1
to Ford Motor Co Ltd, Dagenham, London, January 1970; to BR Swindon Works for rebuild from 19th May 1977 to 8th November 1977; returned to Dagenham.							
D2270	DC RSH	2614 7912	1957	55B	2/68	F	D9
to Briton Ferry Steel Co Ltd, Glamorgan, July 1968; sold via R.E. Trem Ltd, Finningley, Doncaster; scrapped September 1979.							
D2271	DC RSH	2615 7913	1958	55F	10/69	P	2271
to C.F. Booth Ltd, Rotherham, May 1970; privately purchased for preservation and moved to Thomas Hill Ltd, Kilnhurst, 27th July 1972; to Midland Railway, Normanton Barracks, Derby, 7th September 1973; to Midland Railway, Butterley, 10th May 1975; to West Somerset Railway, 15th May 1982.							

BR number	Builder	Works number	Year built	Last BR Shed	Date wdn	P/F	Title
D2272	DC RSH	2616 7914	1957	55F	10/70	P	D2272/ALFIE

to British Fuel Company, Blackburn Coal Concentration Depot, March 1971.

SECTION 4

BR number	Builder	Works number	Year built	Last BR Shed	Date wdn	P/F	Title
D2274	DC RSH	2620 7918	1959	8J	5/69	F	D2274/No.17

to NCB Maltby Colliery, near Rotherham, June 1969; scrapped September 1980.

D2276	DC RSH	2622 7920	1959	30A	8/69	F	D2276

to A.R. Adams Ltd, Newport, Monmouthshire, July 1970; used for spares and remains scrapped in May 1977.

D2279	DC RSH	2656 8097	1960	30E	5/71	P	D2279/No.2

to CEGB Rye House Power Station, Hoddesdon, Hertfordshire, October 1971; to Stour Valley Railway Preservation Society, Essex, about March 1981.

D2280	DC RSH	2657 8098	1960	30E	3/71	P	2/P1381C

to Ford Motor Co Ltd, Dagenham, London, June 1971; to BR Swindon Works for rebuild from 8th July 1977 to 25th November 1977; returned to Dagenham.

D2281	DC RSH	2658 8099	1960	30E	10/68	F	D2281

to Briton Ferry Steel Co Ltd, Glamorgan, April 1969; sold via R.E. Trem Ltd, Finningley, Doncaster; scrapped August 1971.

D2284	DC RSH	2661 8102	1960	30E	4/71	P	D2284

to NCB North Gawber Colliery, Mapplewell, Barnsley, July 1971; to Grimethorpe Colliery, January 1976; to Woolley Colliery, March 1978; to South Yorkshire Railway Preservation Society, Chapeltown, 2nd August 1985; to SYRPS Sheffield, December 1986.

D2294	DC RSH	2674 8127	1960	70D	2/71	F	01

to Shipbreakers (Queenborough) Ltd, Kent, March 1972; scrapped October 1985.

D2298	DC RSH	2679 8157	1960	52A	12/68	P	LORD WENLOCK

to Derwent Valley Light Railway, Layerthorpe, York, April 1969; to Quainton Railway Society, Buckinghamshire, 22nd October 1982.

D2299	DC RSH	2680 8158	1960	52A	1/70	F	D2299/JONAH

to NCB Bestwood Colliery, Nottinghamshire, 9th July 1970; to Hucknall Colliery, 7th August 1970; to Calverton Colliery, May 1978; to Hucknall Colliery, August 1978; to C.F. Booth Ltd, Rotherham, February 1984; scrapped.

D2300	DC RSH	2681 8159	1960	8J	5/69	F	D2300

to NCB Shireoaks Colliery, 25th June 1969; to Steetley Colliery, on loan, 12th September 1974 to 18th November 1974; returned to Shireoaks; to Manton Colliery, 18th October 1978; scrapped August 1986.

D2302	DC RSH	2683 8161	1960	16C	6/69	P	D2302

to British Sugar Corporation, Woodston Factory, Peterborough, August 1969; to BSC Allscott Factory, Shropshire, October 1969; to G.G. Papworth Ltd, Ely, 12th July 1983.

BR number	*Builder*	*Works number*	*Year built*	*Last BR Shed*	*Date wdn*	*P/F*	*Title*
D2304	DC RSH	2685 8163	1960	51A	2/68	F	D8
to Llanelli Steel Co Ltd, Carmarthenshire, July 1968; sold via R.E. Trem Ltd, Finningley, Doncaster; scrapped May 1977.							
D2305	DC RSH	2686 8164	1960	51A	2/68	F	D9
to Llanelli Steel Co Ltd, Carmarthenshire, July 1968; sold via R.E. Trem Ltd, Finningley, Doncaster; scrapped about September 1981.							
D2306	DC RSH	2687 8165	1960	51L	2/68	F	D6
to Llanelli Steel Co Ltd, Carmarthenshire, July 1968; sold via R.E. Trem Ltd, Finningley, Doncaster; scrapped about September 1981.							
D2307	DC RSH	2688 8166	1960	51L	2/68	F	D7
to Llanelli Steel Co Ltd, Carmarthenshire, July 1968; sold via R.E. Trem Ltd, Finningley, Doncaster; scrapped October 1979.							
D2308	DC RSH	2689 8167	1960	51A	2/68	F	D8
to Briton Ferry Steel Co Ltd, Glamorgan, July 1968; sold via R.E. Trem Ltd, Finningley, Doncaster; to Duport Steel Works Ltd, Llanelli, 25th October 1979; scrapped May 1980.							
D2310	DC RSH	2691 8169	1960	52A	1/69	P	D2310
to Coal Mechanisation Ltd, Tolworth, London, April to June 1969.							
D2317	DC RSH	2698 8176	1960	52A	8/69	F	No.10
to NCB Manvers Coal Preparation Plant, Wath-on-Dearne, Rotherham, December 1969; to Cortonwood Colliery, Wombwell, May 1970; scrapped on site by Wath Skip Hire Ltd, 1986.							
D2322	DC RSH	2703 8181	1961	52A	8/68	F	D2322/No.24
to NCB Orgreave Colliery, February 1969; to Treeton Colliery, 1972; to Orgreave Colliery, about January 1973; to Treeton Colliery, about October 1973; to Orgreave Colliery, by April 1974; to Kiveton Park Colliery, 29th April 1980; scrapped Autumn 1985.							
D2324	DC RSH	2705 8183	1961	55B	7/68	P	NPT
to G.W. Talbot Ltd, Coal Concentration Depot, Aylesbury, January 1969.							
D2325	DC RSH	2706 8184	1961	50D	7/68	P	D2325
to NCB Norwich Coal Concentration Depot, December 1968; to Tannick Commercial Repairs, Norwich, October 1986; to John Jolly, Bridgewick Farm, Dengie, Southminster, Essex, March 1987.							
D2326	DC RSH	2707 8185	1961	52A	8/68	F	D2326
to NCB Manvers Coal Preparation Plant, Wath-on-Dearne, Rotherham, February 1969; used for spares; remains scrapped Autumn 1975.							
D2327	DC RSH	2708 8186	1961	52A	8/68	F	No.12/521–12
to NCB Manton Main Colliery, 1969; to Dinnington Colliery, 9th August 1971; to Elsecar Workshops, 3rd May 1973; returned to Dinnington; to Elsecar Workshops, 15th November 1974; to Dinnington Colliery, 20th January 1975; to Coopers (Metals) Ltd, Sheffield, 5th January 1984; scrapped February 1984.							

<table>
<tr><th>BR number</th><th>Builder</th><th>Works number</th><th>Year built</th><th>Last BR Shed</th><th>Date wdn</th><th>P/F</th><th>Title</th></tr>
<tr><td>D2328</td><td>DC
RSH</td><td>2709
8187</td><td>1961</td><td>52A</td><td>9/68</td><td>F</td><td>No.31</td></tr>
<tr><td colspan="8">to NCB Dinnington Colliery, 6th June 1969; to Steetley Colliery, April 1973; to BR Doncaster Works for wheel turning 2nd to 10th February 1977; returned to Steetley; to Shireoaks Colliery by April 1982; to Kiveton Park Colliery, 13th May 1982; to Cortonwood Colliery, 18th July 1985; scrapped on site by Wath Skip Hire Ltd, 1986.</td></tr>
<tr><td>D2329</td><td>DC
RSH</td><td>2710
8188</td><td>1961</td><td>52A</td><td>7/68</td><td>F</td><td>D2329</td></tr>
<tr><td colspan="8">to Derwent Valley Light Railway, Layerthorpe, York, January 1969; sold via Peter Wood & Co Ltd, Eckington, Sheffield; usable spares retained and the remainder scrapped, April 1970.</td></tr>
<tr><td>D2332</td><td>DC
RSH</td><td>2713
8191</td><td>1961</td><td>52A</td><td>6/69</td><td>F</td><td>D2332/LLOYD</td></tr>
<tr><td colspan="8">to Manvers Coal Preparation Plant, Wath-on-Dearne, Rotherham, January 1970; to Cadeby Colliery, 28th August 1975; to Thurcroft Colliery, 14th June 1976; to Shireoaks Colliery, 29th June 1981; to Thurcroft Colliery, 3rd September 1982; to Dinnington Colliery, September 1985; scrapped July 1986.</td></tr>
<tr><td>D2333</td><td>DC
RSH</td><td>2714
8192</td><td>1961</td><td>52A</td><td>9/69</td><td>P</td><td>3/P1062C</td></tr>
<tr><td colspan="8">to Ford Motor Co Ltd, Dagenham, London, December 1969; to BR Swindon Works for rebuild, May to December 1977; returned to Dagenham.</td></tr>
<tr><td>D2334</td><td>DC
RSH</td><td>2715
8193</td><td>1961</td><td>51A</td><td>7/68</td><td>P</td><td>33</td></tr>
<tr><td colspan="8">to NCB Manvers Coal Preparation Plant, Wath-on-Dearne, Rotherham, June 1969; to Thurcroft Colliery, 8th October 1969; to Dinnington Colliery, 19th June 1985; to Maltby Colliery, 24th February 1986.</td></tr>
<tr><td>D2335</td><td>DC
RSH</td><td>2716
8194</td><td>1961</td><td>51A</td><td>7/68</td><td>F</td><td>2</td></tr>
<tr><td colspan="8">to NCB Manvers Coal Preparation Plant, Wath-on-Dearne, Rotherham, June 1969; to Maltby Colliery, about September 1969; scrapped February 1980.</td></tr>
<tr><td>D2336</td><td>DC
RSH</td><td>2717
8195</td><td>1961</td><td>51A</td><td>7/68</td><td>F</td><td>D2336</td></tr>
<tr><td colspan="8">to NCB Manvers Coal Preparation Plant, Wath-on-Dearne, Rotherham, June 1969; used for spares; remains scrapped about February 1978.</td></tr>
<tr><td>D2337</td><td>DC
RSH</td><td>2718
8196</td><td>1961</td><td>51A</td><td>7/68</td><td>P</td><td>D2337/DOROTHY</td></tr>
<tr><td colspan="8">to NCB Manvers Coal Preparation Plant, Wath-on-Dearne, Rotherham, June 1969; to Barnburgh Colliery, June 1974; to Manvers, February 1977.</td></tr>
<tr><td>D2340</td><td>DC
RSH</td><td>2593
7870</td><td>1956</td><td>55F</td><td>10/68</td><td>F</td><td>D1</td></tr>
<tr><td colspan="8">to Briton Ferry Steel Co Ltd, Glamorgan, April 1969; scrapped September 1979.</td></tr>
<tr><td colspan="8">SECTION 5</td></tr>
<tr><td>D2420
06003</td><td>AB</td><td>435</td><td>1959</td><td>RSD</td><td>1984</td><td>P</td><td>06003</td></tr>
<tr><td colspan="8">withdrawn as BR Departmental loco 97804 at Reading Signal Depot; to C.F. Booth Ltd, Rotherham, 25th September 1986; to South Yorkshire Railway Preservation Society, Sheffield, 9th March 1987.</td></tr>
<tr><td>D2432</td><td>AB</td><td>459</td><td>1960</td><td>65A</td><td>12/68</td><td>F</td><td>NFT</td></tr>
<tr><td colspan="8">to Shipbreakers (Queenborough) Ltd, Kent, May 1969; exported to Italy, March 1977 (see Appendix C).</td></tr>
</table>

SECTION 1. D2023 shunts the Middlesbrough Docks of Tees & Hartlepool Port Authority, on 25th June 1979. (A.J. Booth)

SECTION 1. D2138 stands outside the loco shed at Pye Hill Colliery on 15th July 1983, with an NCB Ruston alongside. (A.J. Booth)

SECTION 1. A coal wagon is shunted by D2184 at Southend Coal Concentration Depot, on 27th May 1986. (John Wade)

SECTION 1. Llanelli Steel Company's D10 (formerly D2195) in action on 10th June 1980. (A.J. Booth)

SECTION 1. The former D2199 shunts MGR wagons at Royston Drift Mine, on 27th August 1983. (A.J. Booth)

SECTION 2. D2209 was one of two locos at Manvers which underwent sex changes. Named ERNEST when photographed on 7th June 1969, it was later renamed TRACEY. (A.J. Booth)

SECTION 3. D2229 became a popular loco at Manton, after its transfer from Brookhouse, and here shunts MGR's under the screens on 21st May 1987. (A.J. Booth)

SECTION 3. The NCBOE's D2258 was photographed at C.F. Booth Ltd's scrapyard on 8th January 1987, a couple of days before it was cut up. (A.J. Booth)

SECTION 4. D2284 can now be seen at the South Yorkshire R.P.S. at Sheffield, but is seen here shunting at Woolley Colliery on 29th September 1982.
(John Wade)

SECTION 4. Number 01 (D2294) outside the shed at Shipbreakers (Queenborough) Ltd, on 13th September 1979, with D2070 in the background.
(A.J. Booth)

SECTION 5. 97804 (06003) at C.F. Booth Ltd's scrapyard on 8th January 1987, before moving for preservation to the Sheffield site of the South Yorkshire R.P.S. (A.J. Booth)

SECTION 6. D2519 stands in the muddy yard at NCB Hatfield Colliery, Doncaster, on 18th June 1978. (A.J. Booth)

SECTION 7. Following its rebuild at Hunslet's, CIDER QUEEN (D2578) was photographed at Bulmer's Cider Factory on 27th May 1974. (A.J. Booth)

SECTION 7. The much travelled D2607 stands outside the loco shed at Steetley Colliery, on 28th August 1979. (A.J. Booth)

SECTION 8. Sporting its Andrew Barclay rebuild plate on the cabside, D2738 was photographed on 29th May 1978 at Killoch Colliery. (A.J. Booth)

SECTION 8. Similarly Barclay-plated D2767, resplendant at the Burmah Oil Company's Stanlow refinery on 14th August 1970. (A.C. Baker)

SECTION 9. Immaculate D2865 shunts at APCM's Kilvington works on 17th April 1981, but sadly was scrapped four years later. (A.J. Booth)

SECTION 9. AY1021 (D2866) stands in the yard at Arnott Young's Dalmuir scrapyard, on 24th October 1984. (John Wade)

SECTION 9. DIANE (D2867) shunts Redland wagons at Barrow-on-Soar works, on 3rd April 1986. (A.J. Booth)

SECTION 10. Hunslet shunter D2950, seen here at Ipswich shed in 1958, was later sold to the Llanelli Steel Co Ltd. (N.E. Preedy)

SECTION 11. The first BR loco sold to industry was D2953, seen at March shed on 11th April 1965, and now preserved at the South Yorkshire R.P.S. Sheffield site. (R.N. Pritchard)

SECTION 11. D2956, seen here at BR Doncaster on 27th May 1966, later went to the Norwich yard of A. King & Sons Ltd. (B. Webb)

SECTION 12. Ruston & Hornsby built D2958 in action at C.F. Booth Ltd's Rotherham scrapyard, on 6th September 1969. (A.J. Booth)

SECTION 13. The first 07 Class Ruston, 07001, beside the loading point at Peakstone's Holderness Limeworks, on 24th April 1984. (A.J. Booth)

SECTION 13. Sporting a distinctive camel emblem on its cabside, 07003 standing disused at British Industrial Sand, Oakamoor, on 27th October 1982.
(John Wade)

SECTION 14. Number 11 DULCOTE (formerly D3002) preserved at the ECC Marsh Mills works, Plympton, on 13th September 1982. (A.J. Booth)

SECTION 14. The Guinness Park Royal Brewery now utilise two 08's named LION and UNICORN. The former (D3020) is seen on 27th May 1986.
(John Wade)

SECTION 14. In a striking new livery, D3059 was photographed on 11th June 1981 at Associated British Maltsters Ltd of Airdrie. (A.J. Booth)

SECTION 15. The only Horwich built shunter sold to industry was 08679, seen here at Royston Drift Mine on Christmas Eve 1977. (A.J. Booth)

SECTION 16. SIMON (D3619) pulls a rake of wagons out of the sidings at Moor Green Colliery, on 6th May 1983. (A.J. Booth)

SECTION 16. Darlington built D4067 standing disused in Nailstone Colliery yard on 10th February 1979, prior to preservation at Loughborough.
(A.J. Booth)

SECTION 17. Clayton built Bo-Bo D8568 at Ribblesdale Cement Ltd in April 1978, following its purchase from Hemel Hempstead Lightweight Concrete Co Ltd.
(John Wade)

SECTION 18. Number 62 (D9515) shunts steel bar stock at the BSC Corby Works, on 8th May 1980. (Jim Wade)

SECTION 18. No.7 (D9518) in Ashington Colliery yard on 20th April 1987, with the last 'Paxman' D9555 standing behind. (A.J. Booth)

SECTION 18. D9523 shunts iron ore wagons past the weigh cabin at Glendon East Quarries, on 3rd August 1979. (A.J. Booth)

SECTION 18. Number 64 (D9549) passing through Corby South Wood on 13th May 1980, prior to overhaul, re-gauging, and export to Spain. (Jim Wade)

SECTION 19. No.6 (12071) standing in the yard at Nantgarw Coking Plant, on 28th May 1986. (John Wade)

SECTION 19. 12088 up against the buffer stops at NCBOE Swalwell Disposal Point, County Durham, on 2nd August 1983. (A.J. Booth)

SECTION 19. Not quite up to BR standards, but the shed at NCBOE Bowers Row suffices to protect 12099 on 11th June 1983. (A.J. Booth)

SECTION 19. 510 (12120) heads the final train of wooden wagons from Carter's Coal Depot, Penshaw North, on 23rd August 1973. (I.S. Carr)

SECTION 20. Ashford built diesel electric 15224 standing outside Snowdown Colliery loco shed on 13th September 1979. (A.J. Booth)

SECTION 21. The former Beeston Sleeper Depot shunter (ED10) pushing 5ft 2½in gauge bogies, carrying a girder section, at the Tinsley Viaduct construction contract, Sheffield, of Cleveland Bridge & Engineering Co Ltd. (S.A. Leleux)

SECTION 22. 1ft 6in gauge ZM32 at BR's Horwich Works on 13th October 1963, and now to be seen at Gloddfa Ganol in North Wales. (T.J. Edgington)

SECTION 23. D0226 hauls a short train, including EVENING STAR, at Oakworth Cutting on the Keighley & Worth Valley Railway, on 4th June 1973. (G.W. Morrison)

SECTION 24. Ruston & Hornsby narrow gauge shunter number 85051 shunts at BR's Chesterton Junction PW Depot, Cambridge, on 27th September 1982.
(Ken Scanes)

APPENDIX A. On hire from A.R. Adams Ltd of Newport, ex-BR D2182 shunts at Glyn Neath Opencast Site on 3rd June 1969. (A.C. Baker)

APPENDIX B. Ex-BR Departmental Shunters number 82 (left) and 87 (right) with number 56 just visible on the extreme left, at Thomson's Stockton yard on 24th October 1978. (J.A. Foster)

APPENDIX C. Fresh from overhaul, D3092, D3094, D3098, and D3100 in BREL Derby yard prior to export to Lamco Mining Co of Liberia. (British Railways)

<table>
<tr><th>BR number</th><th>Builder</th><th>Works number</th><th>Year built</th><th>Last BR Shed</th><th>Date wdn</th><th>P/F</th><th>Title</th></tr>
<tr><td colspan="8">SECTION 6</td></tr>
<tr><td>D2511</td><td>HC</td><td>D1202</td><td>1961</td><td>12C</td><td>12/67</td><td>P</td><td>D2511/BRM5477</td></tr>
<tr><td colspan="8">to NCB Brodsworth Colliery, Doncaster, May 1968; to Keighley & Worth Valley Railway, 8th October 1977.</td></tr>
<tr><td>D2513</td><td>HC</td><td>D1204</td><td>1961</td><td>12C</td><td>8/67</td><td>F</td><td>D2513</td></tr>
<tr><td colspan="8">to NCB Cadeby Colliery, Conisbrough, December 1968; scrapped about October 1975.</td></tr>
<tr><td>D2518</td><td>HC</td><td>D1209</td><td>1962</td><td>5A</td><td>2/67</td><td>F</td><td>D2518</td></tr>
<tr><td colspan="8">to NCB Hatfield Colliery, Doncaster, August 1967; later used for spares; remains scrapped June 1973.</td></tr>
<tr><td>D2519</td><td>HC</td><td>D1210</td><td>1962</td><td>5A</td><td>7/67</td><td>F</td><td>D2519</td></tr>
<tr><td colspan="8">to NCB Hatfield Colliery, Doncaster, February 1968; to Keighley & Worth Valley Railway, 3rd April 1982; to Marple & Gillott Ltd, Sheffield, 27th March 1985; scrapped April 1985.</td></tr>
<tr><td colspan="8">SECTION 7</td></tr>
<tr><td>D2554</td><td>HE</td><td>4870</td><td>1956</td><td>70H</td><td>9/83</td><td>P</td><td>97803</td></tr>
<tr><td>05001</td><td colspan="7">withdrawn as BR Departmental loco 97803 on Isle of Wight; to Isle of Wight Steam Railway, Haven Street, 27th August 1984.</td></tr>
<tr><td>D2561</td><td>HE</td><td>4999</td><td>1957</td><td>8F</td><td>8/67</td><td>F</td><td>D3</td></tr>
<tr><td colspan="8">to Llanelli Steel Co Ltd, Carmarthenshire, March 1968; scrapped October 1972.</td></tr>
<tr><td>D2568</td><td>HE</td><td>5006</td><td>1957</td><td>8F</td><td>8/67</td><td>F</td><td>D2568</td></tr>
<tr><td colspan="8">to Briton Ferry Steel Co Ltd, Glamorgan, May 1968; scrapped about May 1969.</td></tr>
<tr><td>D2569</td><td>HE</td><td>5007</td><td>1957</td><td>8F</td><td>8/67</td><td>F</td><td>D6</td></tr>
<tr><td colspan="8">to Briton Ferry Steel Co Ltd, Glamorgan, May 1968; despatched from 8C; scrapped about May 1969.</td></tr>
<tr><td>D2570</td><td>HE</td><td>5008</td><td>1957</td><td>8F</td><td>7/67</td><td>F</td><td>NFT</td></tr>
<tr><td colspan="8">to Briton Ferry Steel Co Ltd, Glamorgan, March 1968; despatched from 8C; scrapped June 1971.</td></tr>
<tr><td>D2578</td><td>HE</td><td>5460</td><td>1958</td><td>62A</td><td>7/67</td><td>P</td><td>2/CIDER QUEEN</td></tr>
<tr><td colspan="8">to Hunslet Engine Co Ltd, Leeds, December 1967; rebuilt as Hunslet 6999; to H.P. Bulmer Ltd, Cider Manufacturers, Hereford, July 1968.</td></tr>
<tr><td>D2587</td><td>HE</td><td>5636</td><td>1959</td><td>62C</td><td>12/67</td><td>P</td><td>2</td></tr>
<tr><td colspan="8">to Hunslet Engine Co Ltd, Leeds, September 1968; rebuilt as Hunslet 7180 with a 384hp engine; to CEGB Chadderton Power Station, September 1969; to Kearsley Power Station, 3rd November 1981; to East Lancs Railway Preservation Society, Bury, March 1983.</td></tr>
<tr><td>D2593</td><td>HE</td><td>5642</td><td>1959</td><td>62C</td><td>12/67</td><td>P</td><td>1</td></tr>
<tr><td colspan="8">to Hunslet Engine Co Ltd, Leeds, January 1969; rebuilt as Hunslet 7179 with a 384hp engine; to CEGB Chadderton Power Station, September 1969; to Kearsley Power Station, 24th September 1981; to East Lancs Railway Preservation Society, Bury, March 1983.</td></tr>
<tr><td>D2598</td><td>HE</td><td>5647</td><td>1960</td><td>50D</td><td>12/67</td><td>F</td><td>SAM</td></tr>
<tr><td colspan="8">to NCB Rossington Colliery, Doncaster, May 1968; to Askern Colliery, July 1971; to Lambton Engine Works, Tyne & Wear, February 1975; scrapped May 1975.</td></tr>
<tr><td>D2599</td><td>HE</td><td>5648</td><td>1960</td><td>50D</td><td>12/67</td><td>F</td><td>F/SE/357</td></tr>
<tr><td colspan="8">to NCB Hickleton Colliery, Doncaster, May 1968; to Frickley Colliery about October 1968; to Askern Colliery, June 1976; scrapped May 1981.</td></tr>
<tr><td>D2600</td><td>HE</td><td>5649</td><td>1960</td><td>50D</td><td>12/67</td><td>F</td><td>D7</td></tr>
<tr><td colspan="8">to Briton Ferry Steel Co Ltd, Glamorgan, April 1968; sold via R.E. Trem Ltd, Finningley, Doncaster; scrapped June 1971.</td></tr>
<tr><td>D2601</td><td>HE</td><td>5650</td><td>1960</td><td>50D</td><td>12/67</td><td>F</td><td>D5</td></tr>
<tr><td colspan="8">to Llanelli Steel Co Ltd, Carmarthenshire, April 1968; sold via R.E. Trem Ltd, Finningley, Doncaster; scrapped 1979.</td></tr>
</table>

<table>
<tr><th>BR number</th><th>Builder</th><th>Works number</th><th>Year built</th><th>Last BR Shed</th><th>Date wdn</th><th>P/F</th><th>Title</th></tr>
<tr><td>D2607</td><td>HE</td><td>5656</td><td>1960</td><td>6G</td><td>12/67</td><td>F</td><td>D2607</td></tr>
<tr><td colspan="8">to NCB Dinnington Colliery, 28th September 1968; to Steetley Colliery about October 1968; to Shireoaks Colliery, on loan, from May to August 1971; returned to Steetley Colliery; to Fence Workshops for overhaul, 28th May 1974; to Steetley Colliery, 30th October 1974; to Shireoaks Colliery, on loan, from 17th May 1975 to July 1975; returned to Steetley Colliery; to Treeton Colliery, on loan, 3rd December 1975; later returned to Steetley Colliery; to BR Doncaster for wheel turning, August 1977; to Steetley Colliery about September 1977; to Shireoaks Colliery, on loan, 26th August 1980; to Steetley Colliery, 16th January 1981; to Coopers (Metals) Ltd, Sheffield, 12th June 1984; scrapped by 4th July 1984.</td></tr>
<tr><td>D2611</td><td>HE</td><td>5660</td><td>1960</td><td>50D</td><td>12/67</td><td>F</td><td>D2611</td></tr>
<tr><td colspan="8">to NCB Yorkshire Main Colliery, Doncaster, May 1968; scrapped about December 1976.</td></tr>
<tr><td>D2613</td><td>HE</td><td>5662</td><td>1960</td><td>50D</td><td>12/67</td><td>F</td><td>D2613/BRM5481</td></tr>
<tr><td colspan="8">to NCB Brodsworth Colliery, Doncaster, May 1968; to Bentley Colliery, 1974; scrapped on site by W. Heselwood Ltd of Sheffield, June 1977.</td></tr>
<tr><td>D2616</td><td>HE</td><td>5665</td><td>1961</td><td>50D</td><td>12/67</td><td>F</td><td>D2616</td></tr>
<tr><td colspan="8">to NCB Hatfield Colliery, Doncaster, May 1968; still there in April 1973 but scrapped later that year.</td></tr>
<tr><td colspan="8">SECTION 8</td></tr>
<tr><td>D2720</td><td>NB</td><td>27815</td><td>1958</td><td>64H</td><td>7/67</td><td>F</td><td>NFT</td></tr>
<tr><td colspan="8">to James N. Connell Ltd, Coatbridge, Lanarkshire, June 1968; scrapped 1971.</td></tr>
<tr><td>D2726</td><td>NB</td><td>27821</td><td>1958</td><td>WW</td><td>2/67</td><td>F</td><td>NFT</td></tr>
<tr><td colspan="8">to Shipbreakers (Queenborough) Ltd, Kent, October 1967; sold via R.E. Trem Ltd, Finningley, Doncaster; scrapped October 1971.</td></tr>
<tr><td>D2736</td><td>NB</td><td>27831</td><td>1958</td><td>65A</td><td>3/67</td><td>F</td><td>D2736</td></tr>
<tr><td colspan="8">to Birds Commercial Motors Ltd, Long Marston, Worcestershire, July 1967; to Birds (Swansea) Ltd, Pontymister Works, Risca, August 1967; to Birds, 40 Acre Site, Cardiff, February 1968; scrapped July 1969.</td></tr>
<tr><td>D2738</td><td>NB</td><td>27833</td><td>1958</td><td>65A</td><td>6/67</td><td>F</td><td>NFT</td></tr>
<tr><td colspan="8">to Andrew Barclay, Sons & Co Ltd, Kilmarnock, October 1967; rebuilt 1968; to NCB Killoch Colliery, Ochiltree, 1969; scrapped 1979.</td></tr>
<tr><td>D2739</td><td>NB</td><td>27834</td><td>1958</td><td>65A</td><td>3/67</td><td>F</td><td>D2739</td></tr>
<tr><td colspan="8">to Birds Commercial Motors Ltd, Long Marston, Worcestershire, July 1967; scrapped September 1969.</td></tr>
<tr><td>D2757</td><td>NB</td><td>28010</td><td>1960</td><td>65A</td><td>7/67</td><td>F</td><td>NFT</td></tr>
<tr><td colspan="8">to Birds (Swansea) Ltd, Pontymister Works, Risca, November 1967; to Birds, 40 Acre Site, Cardiff, February 1968; scrapped October 1970.</td></tr>
<tr><td>D2763</td><td>NB</td><td>28016</td><td>1960</td><td>65A</td><td>6/67</td><td>F</td><td>NFT</td></tr>
<tr><td colspan="8">to Andrew Barclay, Sons & Co Ltd, Kilmarnock, October 1967; rebuilt 1968; to BSC Landore Foundry, Swansea, 1969; scrapped April 1977.</td></tr>
<tr><td>D2767</td><td>NB</td><td>28020</td><td>1960</td><td>65A</td><td>6/67</td><td>P</td><td>NPT</td></tr>
<tr><td colspan="8">to Andrew Barclay, Sons & Co Ltd, Kilmarnock, October 1967; rebuilt 1968; to Burmah Oil Co Ltd, Stanlow, Cheshire, 1969; to East Lancs Railway Preservation Society, Bury, 12th June 1983.</td></tr>
<tr><td>D2774</td><td>NB</td><td>28027</td><td>1960</td><td>65A</td><td>6/67</td><td>P</td><td>D2774</td></tr>
<tr><td colspan="8">to Andrew Barclay, Sons & Co Ltd, Kilmarnock, October 1967; rebuilt 1968; to NCB Killoch Colliery, on hire, 1969 to 1970; sold to NCB; to Celynen North Colliery, Newbridge, March 1971; to BR Canton Depot for repairs, March 1976; to Celynen North Colliery, April 1976; to Celynen South Colliery, Abercarn, about January 1977; to BR Canton Depot for wheel turning, 17th May 1982; to NCB Mountain Ash Works, 27th May 1982; to Celynen South Colliery, 1983; to East Lancs Railway Preservation Society, Bury, September 1986.</td></tr>
</table>

BR number	*Builder*	*Works number*	*Year built*	*Last BR Shed*	*Date wdn*	*P/F*	*Title*
D2777	NB	28030	1960	65A	3/67	F	D2777
to Birds Commercial Motors Ltd, Long Marston, Worcestershire, July 1967; to Birds (Swansea) Ltd, Pontymister Works, Risca, November 1967; scrapped 1968.							

SECTION 9

BR number	*Builder*	*Works number*	*Year built*	*Last BR Shed*	*Date wdn*	*P/F*	*Title*
D2853	YE	2812	1960	8J	6/75	P	PETER
02003	to L.C.P. Fuels Ltd, Shut End, West Midlands, 19th November 1975.						
D2854	YE	2813	1960	8J	2/70	P	D2854
to C.F. Booth Ltd, Rotherham, September 1970.							
D2856	YE	2815	1960	8J	6/75	F	02004
02004	to Redland Roadstone Ltd, Mountsorrel, Leicestershire, 25th September 1975; used for spares; remains, minus engine, to Budden Wood Quarry, Leicestershire; scrapped on site by Vic Berry Ltd of Leicester, 1986.						
D2857	YE	2816	1960	8J	4/71	F	SIR ALFRED NPT
to Birds Commercial Motors Ltd, Long Marston, Worcestershire, November 1971.							
D2858	YE	2817	1960	9A	2/70	P	NPT
to Hutchinson Estate & Dock Co (Widnes) Ltd, Widnes, August 1970; to Fisons Fertilisers, Widnes, September 1978; to Lowton Metals Ltd, Haydock, 5th March 1981; to Butterley Ltd, Ripley, November 1986.							
D2860	YE	2843	1961	8J	12/70	P	D2860
to Curator of Historical Relics, BR, Preston Park, Brighton, March 1973; to National Railway Museum, York, 26th November 1977; to Thomas Hill (Rotherham) Ltd, Kilnhurst, for overhaul and repaint, 14th September 1978; to National Railway Museum, York, 3rd January 1979.							
D2862	YE	2845	1961	10D	12/69	F	ND3/63000359
to Tilsley & Lovatt Ltd, Trentham, Staffordshire, March 1970; overhauled and sold to NCB; to Norton Colliery, January 1971; to Chatterley Whitfield Colliery, September 1971; to Norton Colliery, October 1971; scrapped about April 1979.							
D2865	YE	2848	1961	50D	3/70	F	NFT
to APCM Kilvington, Nottinghamshire, September 1970; to Blue Circle, Beeston Depot, Nottingham, for storage, 1984; to Vic Berry Ltd, Leicester, January 1985; scrapped May 1985.							
D2866	YE	2849	1961	9A	2/70	P	AY1021
to Arnott Young Ltd, Dalmuir, Dunbartonshire, June 1970; to BREL, Glasgow, for repairs, April 1977; returned to Dalmuir.							
D2867	YE	2850	1961	6A	9/70	P	DIANE
to Redland Roadstone Ltd, Mountsorrel, Leicestershire, October 1970; to Barrow-on-Soar Works, Leicestershire, late 1979.							
D2868	YE	2851	1961	10D	12/69	P	SAM
to Lunt, Comley & Pitt Ltd, Shut End, Staffordshire, October 1970.							

SECTION 10

BR number	*Builder*	*Works number*	*Year built*	*Last BR Shed*	*Date wdn*	*P/F*	*Title*
D2950	HE	4625	1954	50D	12/67	F	D4
to Llanelli Steel Co Ltd, Carmarthenshire, April 1968; sold via R.E. Trem Ltd, Finningley, Doncaster; purchased for preservation about May 1980 and stored at Thyssen Ltd, Old Castle Depot, Llanelli; scrapped about 1983.							

SECTION 11

BR number	*Builder*	*Works number*	*Year built*	*Last BR Shed*	*Date wdn*	*P/F*	*Title*
D2953	AB	395	1955	30A	6/66	P	D2953
to Thames Matex Ltd, West Thurrock, Essex, June 1966; loaned to BP Refinery (Kent) Ltd, 1967; loaned to Shell Mex & BP Ltd, Purfleet, at various times; to South Yorkshire Railway Preservation Society, Chapeltown, 15th December 1985; to SYRPS, Sheffield, December 1986.							

BR number	*Builder*	*Works number*	*Year built*	*Last BR Shed*	*Date wdn*	*P/F*	*Title*
D2956	AB	398	1956	36A	5/66	P	D2956

to A. King & Sons Ltd, Norwich, July 1966; to A. King & Sons Ltd, Snailwell, Cambridgeshire, about September 1981; to East Lancs Railway Preservation Society, Bury, 30th July 1985.

D2956	AB	424	1958	36A	11/67	F	D5

The second D2956; withdrawn as BR Departmental loco number 81; to Briton Ferry Steel Co Ltd, Glamorgan, March 1968; scrapped August 1969.

SECTION 12

D2958	RH	390777	1956	30A	1/68	F	NFT

to C.F. Booth Ltd, Rotherham, May 1968; sold via R.E. Trem Ltd, Finningley, Doncaster; scrapped October 1984.

SECTION 13

D2985	RH	480686	1962	70D	7/77	P	NPT

07001 to Tilsley & Lovatt Ltd, Trentham, Staffordshire, for overhaul, 1978; to Peakstone Ltd, Holderness Limeworks, Peak Dale, Derbyshire, May 1978.

D2986	RH	480687	1962	70D	7/77	F	NFT

07002 to Powell Duffryn Fuels Ltd, NCBOE Coed Bach Disposal Point, Kidwelly, Dyfed, April 1978; scrapped late 1982.

D2987	RH	480688	1962	70D	10/76	F	07003

07003 to R.E. Trem Ltd, Finningley, Doncaster, March 1977; to British Industrial Sand Ltd, Oakamoor, Staffordshire, about October 1978; scrapped May 1985.

D2989	RH	480690	1962	70D	7/77	P	LANGBAURGH

07005 to Resco (Railways) Ltd, Woolwich, for overhaul, June 1978; to ICI, Wilton Works, Middlesbrough, April 1979.

D2990	RH	480691	1962	70D	7/77	F	NFT

07006 to Powell Duffryn Fuels Ltd, NCBOE Coed Bach Disposal Point, Kidwelly, Dyfed, April 1978; scrapped on site by T. Davis of Llanelli, October 1984.

D2994	RH	480695	1962	70F	10/76	P	D2994

07010 to Winchester & Alton Railway, New Alresford, Hampshire, August 1978; to West Somerset Railway, 19th May 1980.

D2995	RH	480696	1962	70D	7/77	P	07011/CLEVELAND

07011 to Resco (Railways) Ltd, Woolwich, for overhaul, June 1978; to ICI Billingham Works, on hire, March 1979; returned to Resco, 12th November 1979; to ICI Wilton Works, Middlesbrough, 4th September 1980.

D2996	RH	480697	1962	70D	7/77	P	07012

07012 to Powell Duffryn Fuels Ltd, NCBOE Cwm Mawr Disposal Point, Tumble, Dyfed, April 1978; to NCBOE Coed Bach Disposal Point, early 1982.

D2997	RH	480698	1962	70D	7/77	P	NPT

07013 to Resco (Railways) Ltd, Woolwich, for overhaul, May 1978; to Dow Chemicals Ltd, Kings Lynn, October 1978.

SECTION 14

D3000	Derby		1952	82A	11/72	P	3000

to NCB Hafodyrynys Colliery, Pontypool, March 1973; to BR Canton Depot for repairs, August 1975; to Hafodyrynys, about November 1975; to Bargoed Colliery, Mid Glamorgan, 17th July 1978; to BR Canton Depot, October 1979; to Bargoed Colliery, October 1979; to BR Canton Depot, June 1980; to Bargoed Colliery, July 1980; to Mountain Ash Colliery, 10th July 1981; to Mardy Colliery, 13th November 1981; to Brighton Works Locomotive Association, 18th March 1987, arriving there 21st March 1987.

BR number	Builder	Works number	Year built	Last BR Shed	Date wdn	P/F	Title
D3002	Derby		1952	82A	7/72	P	11/DULCOTE
	to Foster Yeoman Ltd, Merehead Stone Terminal, Somerset, November 1972; to BR Bath Road Depot for repairs, 13th April 1976; to Merehead, June 1976; to Plym Valley Railway Association, May 1982. and stored at ECC Marsh Mills Works, Plympton.						
D3003	Derby		1952	82A	7/72	P	22/MEREHEAD
	to Foster Yeoman Ltd, Merehead Stone Terminal, Somerset, May 1973; to BREL Derby Works, for repairs, 30th June 1974; returned to Merehead; to Wanstrow Childrens Playground, about February 1982.						
D3011	Derby		1952	70D	10/72	F	LICKEY
	to British Leyland Ltd, Longbridge, Birmingham, January 1973; to BR Tyseley Depot for wheel turning, November 1981; returned to Longbridge; to Marple & Gillott Ltd, Sheffield, 6th December 1985; scrapped December 1985.						
D3014	Derby		1952	70D	10/72	P	D3014
	to NCB Merthyr Vale Colliery, Aberfan, September 1973; to BR Canton Depot, December 1974; to Merthyr Vale Colliery, about January 1975; to BR Canton Depot, February 1980; to Merthyr Vale Colliery, February 1980; to BR Canton Depot, September 1980; to Merthyr Vale Colliery, October 1980; to BR Canton Depot, 3rd October 1981; to Merthyr Vale Colliery, 27th December 1981.						
D3019	Derby		1953	8J	7/73	P	D3019/GWYNETH
	to Powell Duffryn Fuels Ltd, NCBOE Gwaun-cae-Gurwen Disposal Point, West Glamorgan, November to December 1973; to BR Canton Depot, 4th June 1978; to Gwaun-cae-Gurwen, 8th December 1978.						
D3022	Derby		1953	41A	9/80	P	08015
08015	to Severn Valley Railway, 27th May 1983; despatched from BREL Swindon Works.						
D3023	Derby		1953	9D	5/80	P	08016
08016	to Hargreaves Industrial Services Ltd, NCBOE British Oak Disposal Point, Crigglestone, West Yorkshire, October 1980.						
D3030	Derby		1953	41A	3/85	P	LION
08022	to Guinness Ltd, Park Royal, London, 20th July 1985; despatched from Swindon Works.						
D3038	Derby		1953	9A	12/72	F	2100/525
	to NCB Ashington Colliery, October 1973; despatched from 9D; to Bates Colliery, Blyth, about March 1974; scrapped 1980.						
D3044	Derby		1954	16A	8/74	P	33/MENDIP
08032	to BR Derby Works for overhaul, August 1974; to Foster Yeoman Ltd, Merehead Stone Terminal, Somerset, February 1975; to BR Gloucester shed for repairs, 10th February 1980; returned to Merehead.						
D3059	Derby		1954	16F	5/80	P	08046
08046	to Associated British Maltsters Ltd, Airdrie, 21st January 1981; despatched from 16C; to BR Motherwell Depot for minor repairs, 7th February 1981; returned to Airdrie; to Caledonian Railway Ltd, Brechin, Tayside, December 1985.						
D3067	Darlington		1953	52A	2/80	P	08054
08054	to Tilcon Ltd, Swinden Limeworks, Grassington, about August 1980.						
D3074	Darlington		1953	40A	6/84	P	UNICORN
08060	to Guinness Ltd, Park Royal, London, 20th July 1985; despatched from Swindon Works.						
D3087	Derby		1954	8F	7/73	F	NFT
	to CEGB Walsall Power Station, October 1973; to BR Tyseley Depot, November 1981; returned to Walsall; scrapped on site by Thos. W. Ward Ltd, May 1983.						
D3088	Derby		1954	2F	12/73	F	D3088/2100–526
	to NCB Ashington Colliery, about May 1974; to Bates Colliery, Blyth, December 1974; to Lambton Engine Works, Philadelphia, April 1979; to Bates Colliery, September 1979; to Lambton Engine Works, 29th June 1981; to Bates Colliery, 15th February 1983; scrapped on site week ending 26th October 1985.						

BR number	Builder	Works number	Year built	Last BR Shed	Date wdn	P/F	Title	
D3099	Derby		1955	73F	10/72	F	NFT	to Shipbreakers (Queenborough) Ltd, Kent, July 1973; used for spares and dismantled; remains gone by May 1981.
D3101	Derby		1955	73F	5/72	P	D3101	to ARC (East Midlands) Ltd, Loughborough, February 1973; to Great Central Railway, Loughborough, about November 1984.
D3102 08077	Derby		1955	31A	11/77	P	08077	to Wiggins Teape & Co Ltd, Fort William, December 1978; to BREL Glasgow for repairs, June 1981; returned to Wiggins Teape.

SECTIONS 15/16

BR number	Builder	Works number	Year built	Last BR Shed	Date wdn	P/F	Title	
D3174 08108	Derby		1955	31A	7/84	P	08108	to Dower Wood & Co Ltd, Newmarket, Suffolk, 1st August 1984.
D3179 08113	Derby		1955	86A	3/84	P	08113	to P.D. Fuels Ltd, NCBOE Gwaun-cae-Gurwen, 6th August 1984.
D3180 08114	Derby		1955	36A	11/83	P	08114	to Gloucester & Warwickshire Railway, Toddington, 3rd October 1984; despatched from BREL Swindon Works.
D3183	Derby		1955	82C	12/72	P	D3183	to NCB Merthyr Vale Colliery, Aberfan, March 1973; to BR Canton Depot for repairs, May 1975; to Merthyr Vale Colliery, June 1975; to BR Canton, September 1980; to Merthyr Vale Colliery, October 1980; to BR Canton, July 1985; to Merthyr Vale Colliery, December 1985.
D3190 08123	Derby		1955	5A	3/84	P	GEORGE MASON	to Wallingford Railway, 7th June 1985; despatched from BREL Swindon Works.
D3201 08133	Derby		1955	40A	9/80	P	08133	to Sheerness Steel Co Ltd, 6th October 1981; despatched from BREL Swindon Works.
D3225 08157	Darlington		1955	73F	4/77	P	08157	to Independent Sea Terminals, Ridham Dock, Kent, 28th July 1977; despatched from BREL Eastleigh Works.
D3255	Derby		1956	85B	12/72	P	3255	to NCB Blaenavon Colliery, March 1973; to Bargoed Colliery, March 1973; to BR Canton Depot for repairs, about September 1975; to Bargoed Colliery, about April 1976; to BR Canton Depot, August 1976; to Bargoed Colliery, 1976; to Canton Depot, May 1977; to Bargoed Colliery, about July 1977; to Canton Depot, March 1978; to Bargoed Colliery, about May 1978; to Canton Depot, 23rd May 1979; to Bargoed Colliery, about August 1979; to Canton Depot, 6th April 1981; to Mardy Colliery, 27th May 1981; to Mountain Ash Colliery, 28th December 1981; to Mardy Colliery, 1982.
D3261	Derby		1956	86A	12/72	P	3261	to NCB Tower Colliery, Hirwaun, Glamorgan, July 1973; to BR Canton Depot for repairs, September 1975; to Tower Colliery, October 1975; to Canton Depot, October 1977; to Tower Colliery, November 1977; to Canton Depot, November 1978; to BR Swindon Works, April 1979; to Tower Colliery, 29th November 1979.
D3286 08216	Derby		1956	16C	11/80	P	08216	to Sheerness Iron & Steel Co Ltd, 12th May 1983; despatched from Swindon Works.
D3308 08238	Darlington		1956	85B	3/84	P	08238	sold to Forest Free Mining, Tetbury, for Parkend Mine in the Forest of Dean, 1984; but stored at Gloucester Depot and still there Summer 1987.
D3336 08266	Darlington		1957	41J	3/85	P	08266	to Keighley & Worth Valley Railway, 21st November 1985; despatched from Swindon Works.

BR number	Builder	Works number	Year built	Last BR Shed	Date wdn	P/F	Title
D3358	Derby		1957	82C	1/83	P	08288
08288	to Winchester & Alton Railway, 21st October 1984; despatched from Swindon Works.						
D3362	Derby		1957	65A	5/84	P	08292
08292	to Deanside Transit Ltd, Glasgow, July 1984.						
D3390	Derby		1957	16A	12/82	P	08320
08320	sold to Forest Free Mining, Tetbury, for Parkend Mine in the Forest of Dean, 2nd October 1984; but stored at Gloucester Depot and still there Summer 1987.						
D3415	Derby		1958	67C	10/83	P	08345
08345	to Deanside Transit Ltd, Glasgow, May 1985.						
D3420	Crewe		1957	86A	1/84	P	08350
08350	to North Staffordshire Railway, 18th September 1984; despatched from Swindon Works.						
D3429	Crewe		1958	86A	1/84	P	08359
08359	to North Staffordshire Railway, 20th September 1984; despatched from Swindon Works; to Peak Rail, Buxton, 11th January 1987.						
D3452	Darlington		1957	16A	6/68	P	D3452/73601
	to ECC Ports Ltd, Fowey, Cornwall, September 1968.						
D3462	Darlington		1957	84A	6/83	P	08377
08377	to Dean Forest Railway, 12th March 1986; despatched from BREL Swindon Works.						
D3476	Darlington		1957	16A	5/68	P	D3476/73603
	to ECC Ports Ltd, Fowey, Cornwall, September 1968.						
D3489	Darlington		1958	16A	4/68	P	COLONEL TOMLINE
	to Felixtowe Dock & Railway Co, Suffolk, August 1968; to BREL Swindon Works for repairs, 30th January 1980; to Felixtowe, 19th May 1980; to BR Stratford Depot for repairs, 11th October 1984; to Felixtowe, December 1984.						
D3497	Doncaster		1957	16B	4/68	P	D3497
	to ECC Ports Ltd, Fowey, Cornwall, August 1968.						
D3513	Derby		1958	82A	7/85	P	NPT
08398	to ECC Ports Ltd, Fowey, Cornwall, 3rd February 1986.						
D3558	Derby		1958	62A	7/85	P	08443
08443	to Scottish Grain Distillers, Cambus Distillery, Alloa, 14th January 1986.						
D3559	Derby		1958	86A	11/86	P	08444
08444	to Bodmin & Wenford Railway, about February, 1987.						
D3586	Crewe		1958	15A	9/85	P	08471
08471	to Severn Valley Railway, 14th April 1986; despatched from BREL Swindon Works.						
D3591	Crewe		1958	62A	9/85	P	08476
08476	to Swanage Railway, 21st March 1986; despatched from BREL Swindon Works.						
D3613	Darlington		1958	40A	2/69	F	DAVID
	to NCB Bestwood Colliery, Nottinghamshire, August 1969; to Linby Colliery, by September 1971; to Moor Green Colliery, November 1971; scrapped on site by Vic Berry Ltd, 10th April 1985.						
D3618	Darlington		1958	40A	4/69	F	ROBIN/D16
	to NCB Bestwood Colliery, August 1969; to Annesley Colliery, March 1970; to BR Toton Depot for repairs, June 1974; to Cotgrave Colliery, July 1980; to Moor Green Colliery, 30th March 1981; scrapped on site by Vic Berry Ltd, 26th March 1985.						
D3619	Darlington		1958	40A	2/69	F	SIMON/D15
	to NCB Gedling Colliery, Nottinghamshire, September 1969; to Bestwood Colliery, January 1970; to Linby Colliery, about July 1971; to Moor Green Colliery, November 1975; scrapped on site by Vic Berry Ltd, week ending 5th April 1985.						

BR number	Builder	Works number	Year built	Last BR Shed	Date wdn	P/F	Title
D3638	Darlington		1958	52A	11/70	F	9185–61
	to NCB Bates Colliery, Blyth, March 1971; to Ashington Central Workshops, March 1975; dismantled for spares, and remains scrapped September 1975.						
D3642	Darlington		1958	36C	6/69	F	37
	to BSC Redbourn Works, Scunthorpe, September 1969; to BSC Appleby-Frodingham Works, Scunthorpe, October 1975; scrapped October 1978.						
D3648	Darlington		1959	52A	1/71	F	9185–60
	to NCB Bates Colliery, Blyth, March 1971; scrapped on site by L. Marley & Co Ltd of Stanley, February 1977.						
D3763	Derby		1959	81A	3/77	P	08596
08596	to Bowaters UK Paper Co Ltd, Sittingbourne, Kent, 16th May 1977; to BR Swindon Works for overhaul, 20th November 1981; returned to Bowaters, 4th January 1982.						
D3765	Derby		1959	5A	11/86	P	08598
08598	to Powell Duffryn Fuels Ltd, NCBOE Gwaun-cae-Gurwen, 16th January 1987.						
D3846	Horwich		1959	8J	6/76	F	08679
08679	to NCB North Gawber Colliery, Mapplewell, Barnsley, June 1976; to Royston Drift Mine, September 1976; to North Gawber Colliery, April 1979; to C.F. Booth Ltd, Rotherham, 18th April 1986; scrapped April 1986.						
D3984	Derby		1960	51L	2/86	P	08816
08816	to Cobra Railfreight Ltd, Middlesbrough, February 1986.						
D4056	Darlington		1961	40B	6/72	F	D4056/55
	to NCB Ashington Colliery, January 1973; to Shilbottle Colliery, about June 1974; scrapped on site by T.J. Thomson Ltd, March 1983.						
D4067	Darlington		1961	41J	12/70	P	D4067/1802-B4
	to NCB Betteshanger Colliery, Kent, April 1971; to Snowdown Colliery, 27th May 1976; to Nailstone Colliery, Leicestershire, 14th June 1976; to BR Doncaster for repairs, October 1976; to Nailstone Colliery, December 1976; to Main Line Steam Trust, Loughborough, 5th February 1980.						
D4068	Darlington		1961	40B	6/72	F	No.56/9300-116
	to NCB Ashington Colliery, January 1973; to Shilbottle Colliery, January 1973; to Lambton Engine Works, Philadelphia, April 1979; to Whittle Colliery, April 1980; scrapped on site by C.F. Booth Ltd of Rotherham, October 1985.						
D4069	Darlington		1961	41J	4/72	P	9300-111
	to NCB Ashington Colliery, September 1972; to Whittle Colliery, October 1972; to Lambton Engine Works, Philadelphia, February 1978; to Whittle Colliery, March 1979.						
D4070	Darlington		1961	41J	4/72	F	No.52/9300-112
	to NCB Ashington Colliery, September 1972; to Shilbottle Colliery, November 1972; to Ashington Colliery, September 1974; to Lambton Engine Works, Philadelphia, October 1975; to Bates Colliery, Blyth, November 1976; to Whittle Colliery, April 1977; to Lambton Engine Works, April 1980; to Whittle Colliery, 5th June 1981; scrapped on site by C.F. Booth Ltd of Rotherham, October 1985.						
D4072	Darlington		1961	31B	4/72	F	No.53/9300-114
	to NCB Ashington Colliery, October 1972; to Whittle Colliery, November 1972; to Lambton Engine Works, Philadelphia, June 1977; to Whittle Colliery, February 1978; to Lambton Engine Works, June 1978; to Whittle Colliery, August 1978; to Lambton Engine Works, September 1978; to Whittle Colliery, November 1978; to Lambton Engine Works, December 1979; to Whittle Colliery, June 1980; to Lambton Engine Works, 14th April 1981; to South Hetton Colliery, 10th May 1982; to Lambton Engine Works, 27th September 1982; to Lambton Engine Works about January 1983; scrapped on site by C.F. Booth Ltd of Rotherham, November 1985.						

<table>
<tr><th>BR number</th><th>Builder</th><th>Works number</th><th>Year built</th><th>Last BR Shed</th><th>Date wdn</th><th>P/F</th><th>Title</th></tr>
<tr><td>D4074</td><td>Darlington</td><td></td><td>1961</td><td>31B</td><td>4/72</td><td>F</td><td>No.54</td></tr>
<tr><td colspan="8">to NCB Ashington Colliery, October 1972; to Whittle Colliery, November 1972; to Lambton Engine Works, Philadelphia, February 1977; scrapped August 1978.</td></tr>
<tr><td colspan="8">SECTION 17</td></tr>
<tr><td>D8568</td><td>CE</td><td>4365U/69</td><td>1963</td><td>66A</td><td>10/71</td><td>P</td><td>D8568</td></tr>
<tr><td colspan="8">to Hemel Hempstead Lightweight Concrete Co Ltd, Cupid Green, Hertfordshire, September 1972; to Ribblesdale Cement Ltd, Clitheroe, Lancashire, from 15th to 22nd June 1977; to North Yorkshire Moors Railway, 11th February 1983.</td></tr>
<tr><td colspan="8">SECTION 18</td></tr>
<tr><td>D9500</td><td>Swindon</td><td></td><td>1964</td><td>86A</td><td>4/69</td><td>P</td><td>9312-92</td></tr>
<tr><td colspan="8">to NCB Ashington Colliery, November 1969; to Lambton Engine Works, Philadelphia, July 1978; to Ashington Colliery, June 1979; to Lambton Engine Works, 12th May 1983; to Ashington Colliery, 25th July 1983.</td></tr>
<tr><td>D9502</td><td>Swindon</td><td></td><td>1964</td><td>86A</td><td>4/69</td><td>P</td><td>D9502/9312-97</td></tr>
<tr><td colspan="8">to NCB Ashington Colliery, July 1969; to Burradon Colliery, about August 1969; to Ashington Central Workshops, about July 1973; to Burradon Colliery, about February 1974; to Backworth Colliery, January 1976; to Weetslade Loco Shed, about June 1976; to Ashington Colliery, 24th August 1981; to Lambton Engine Works, 28th April 1983; to Ashington Colliery, 12th May 1983.</td></tr>
<tr><td>D9503</td><td>Swindon</td><td></td><td>1964</td><td>50B</td><td>4/68</td><td>F</td><td>65/8411-25</td></tr>
<tr><td colspan="8">to Stewarts & Lloyds Minerals Ltd, Harlaxton Quarries, Lincolnshire, November 1968; to Corby Quarries, July 1974; scrapped September 1980.</td></tr>
<tr><td>D9504</td><td>Swindon</td><td></td><td>1964</td><td>50B</td><td>4/68</td><td>P</td><td>506/2233-506</td></tr>
<tr><td colspan="8">to NCB Philadelphia Loco Shed, County Durham, November 1968; to Bolden Colliery, August 1973; to Philadelphia, September 1973; to Bolden Colliery, February 1974; to Backworth Colliery, December 1974; to Burradon Colliery, January 1975; to Weetslade Loco Shed, January 1976; to Lambton Engine Works, Philadelphia, 21st April 1981; to Ashington Colliery, 11th September 1981.</td></tr>
<tr><td>D9505</td><td>Swindon</td><td></td><td>1964</td><td>50B</td><td>4/68</td><td>F</td><td>MICHLOW</td></tr>
<tr><td colspan="8">to APCM, Hope, Derbyshire, September 1968; sold for export (see Appendix C); left Hope 5th May 1975.</td></tr>
<tr><td>D9507</td><td>Swindon</td><td></td><td>1964</td><td>50B</td><td>4/68</td><td>F</td><td>55/8311-35</td></tr>
<tr><td colspan="8">to Stewarts & Lloyds Minerals Ltd, Corby Quarries, November 1968; to BSC Steelworks Disposal Site, Corby, December 1980; scrapped on site by Shanks & McEwan of Corby, September 1982.</td></tr>
<tr><td>D9508</td><td>Swindon</td><td></td><td>1964</td><td>87A</td><td>10/68</td><td>F</td><td>No.9/9312-99</td></tr>
<tr><td colspan="8">to NCB Ashington Colliery, March 1969; despatched from 86A; withdrawn November 1983; scrapped on site by D. Short Ltd of North Shields, 17th January 1984.</td></tr>
<tr><td>D9510</td><td>Swindon</td><td></td><td>1964</td><td>50B</td><td>4/68</td><td>F</td><td>60/8411-23</td></tr>
<tr><td colspan="8">to Stewarts & Lloyds Minerals Ltd, Buckminster Quarries, Lincolnshire, December 1968; to Corby Quarries, June 1972; to BSC Tube Works, Corby, January 1981; scrapped on site by Shanks & McEwan of Corby, August 1982.</td></tr>
<tr><td>D9511</td><td>Swindon</td><td></td><td>1964</td><td>50B</td><td>4/68</td><td>F</td><td>9312-98</td></tr>
<tr><td colspan="8">to NCB Ashington Colliery, November 1968; to Bates Colliery, Blyth, May 1969; to Burradon Colliery, May 1969; to Ashington Colliery, about October 1972; dismantled for spares after a fire; remains scrapped July 1979.</td></tr>
<tr><td>D9512</td><td>Swindon</td><td></td><td>1964</td><td>50B</td><td>4/68</td><td>F</td><td>63/8411-24</td></tr>
<tr><td colspan="8">to Stewarts & Lloyds Minerals Ltd, Buckminster Quarries, Lincolnshire, December 1968; to Corby Quarries, September 1972; used for spares; to BSC Steelworks Disposal Site, Corby, December 1980; scrapped about February 1982.</td></tr>
</table>

BR number	Builder	Works number	Year built	Last BR Shed	Date wdn	P/F	Title
D9513	Swindon		1964	86A	3/68	P	38/2100-524
to Arnott Young Ltd, Parkgate, Rotherham, July 1968; despatched from 85A; to Hargreaves Industrial Services Ltd, NCBOE British Oak Disposal Point, Crigglestone, November 1968; to NCBOE Bowers Row Disposal Point, Astley, September 1969; sold to NCB North East Area; to Allerton Bywater Central Workshops, West Yorkshire, about October 1973; to Ashington Colliery, January 1974; to Backworth Colliery, July 1974; to Burradon Colliery, July 1974; to Backworth Colliery, January 1976; to Lambton Engine Works, Philadelphia, November 1976; to Ashington Colliery, February 1977; to Lambton Engine Works, September 1977; to Ashington Colliery, November 1977; to Lambton Engine Works, 30th June 1982; to Ashington Colliery, 17th February 1983.							
D9514	Swindon		1964	86A	4/69	F	No.4/9312-96
to NCB Ashington Colliery, July 1969; to BR Gosforth Depot for repairs, October 1975; to Ashington Colliery, November 1975; to Lambton Engine Works, Philadelphia, about August 1977; to Ashington Colliery, about 1978; scrapped on site, November 1985.							
D9515	Swindon		1964	50B	4/68	F	62/8411-22
to Stewarts & Lloyds Minerals Ltd, Buckminster Quarries, Lincolnshire, 2nd November 1968; to Corby Quarries, September 1972; to BSC Steelworks Disposal Site, Corby, December 1980; to Hunslet Engine Co Ltd, Leeds, December 1981; overhauled and converted to 5ft 6in gauge; exported to Spain from Goole Docks, June 1982.							
D9516	Swindon		1964	50B	4/68	P	D9516
to Stewarts & Lloyds Minerals Ltd, Corby Quarries, November 1968; to BSC Steelworks Disposal Site, Corby, December 1980; to Great Central Railway, Loughborough, 17th October 1981.							
D9517	Swindon		1964	86A	10/68	F	No.8/9312-93
to NCB Ashington Colliery, November 1969; to Lambton Engine Works, Philadelphia, June 1977; to Ashington Colliery, September 1977; withdrawn November 1983; scrapped on site by D. Short Ltd of North Shields, January 1984.							
D9518	Swindon		1964	86A	4/69	P	No.7/9312-95
to NCB Ashington Colliery, June to July 1969; to Lambton Engine Works, Philadelphia, June 1975; to Ashington Colliery by June 1976; to Lambton Engine Works, 5th September 1980; to Ashington Colliery, 3rd December 1980.							
D9520	Swindon		1964	50B	4/68	P	45/8311-24
to Stewarts & Lloyds Minerals Ltd, Glendon Quarries, Northamptonshire, December 1968; to Corby Quarries, January 1970; to BSC Tube Works, Corby, October 1980; to North Yorkshire Moors Railway, Grosmont, 16th March 1981; to Cottesmore Museum, 21st February 1984.							
D9521	Swindon		1964	87A	4/69	P	No.3/9312-90
to NCB Ashington Colliery, March 1970; to BR Gosforth Depot for repairs, March 1976; to Ashington Colliery, about May 1976; to Lambton Engine Works, Philadelphia, November 1977; to Ashington Colliery, 1978; to Lambton Engine Works, 17th January 1982; to Ashington Colliery, 30th June 1982; to Lambton Engine Works, March 1983; to Ashington Colliery, March 1983.							
D9523	Swindon		1964	50B	4/68	P	D9523
to Stewarts & Lloyds Minerals Ltd, Glendon Quarries, Northamptonshire, December 1968; to Corby Quarries, 28th May 1980; to BSC Steelworks Disposal Site, December 1980; to Great Central Railway, Loughborough, 17th October 1981.							
D9524	Swindon		1964	87A	4/69	P	No.8/144-8
to BP Refinery (Grangemouth) Ltd, Grangemouth, July 1970; re-engined by Andrew Barclay's with a Dorman 500hp engine; to BR Eastfield Depot for repairs, March 1978; returned to BP; to Scottish Railway Preservation Society, Falkirk, 9th September 1981.							

<table>
<tr><th>BR number</th><th>Builder</th><th>Works number</th><th>Year built</th><th>Last BR Shed</th><th>Date wdn</th><th>P/F</th><th>Title</th></tr>
<tr><td>D9525</td><td>Swindon</td><td></td><td>1964</td><td>50B</td><td>4/68</td><td>P</td><td>507/2233-507</td></tr>
<tr><td colspan="8">to NCB Philadelphia Loco Shed, County Durham, November 1968; to Burradon Colliery, March 1975; to Ashington Colliery, March 1975; to Backworth Colliery, December 1975; to Weetslade Washer, January 1981; to Ashington Colliery, 24th April 1981; to Lambton Engine Works, 25th July 1983; to Ashington Colliery, 19th December 1983.</td></tr>
<tr><td>D9526</td><td>Swindon</td><td></td><td>1964</td><td>86A</td><td>11/68</td><td>P</td><td>D9526</td></tr>
<tr><td colspan="8">to APCM, Westbury, Wiltshire, January 1970; to APCM, Dunstable, 28th May 1971; to APCM, Westbury, 24th November 1971; to West Somerset Railway, 3rd April 1980.</td></tr>
<tr><td>D9527</td><td>Swindon</td><td></td><td>1965</td><td>86A</td><td>4/69</td><td>F</td><td>No.6/9312-94</td></tr>
<tr><td colspan="8">to NCB Ashington Colliery, July 1969; to Lambton Engine Works, Philadelphia, May 1977; to Ashington Colliery, September 1977; to Lambton Engine Works, September 1978; to Ashington Colliery, about October 1978; withdrawn November 1983; scrapped on site by D. Short Ltd of North Shields, January 1984.</td></tr>
<tr><td>D9528</td><td>Swindon</td><td></td><td>1965</td><td>86A</td><td>3/69</td><td>F</td><td>No.2/9312-100</td></tr>
<tr><td colspan="8">to NCB Ashington Colliery, March 1969; to BR Gosforth Depot for repairs, March 1976; to Ashington Colliery by May 1976; to Lambton Engine Works, Philadelphia, June 1977; to Ashington Colliery by November 1977; scrapped December 1981.</td></tr>
<tr><td>D9529</td><td>Swindon</td><td></td><td>1965</td><td>50B</td><td>4/68</td><td>P</td><td>61/8411-20</td></tr>
<tr><td colspan="8">to Stewarts & Lloyds Minerals Ltd, Buckminster Quarries, Lincolnshire, August 1968; to Corby Quarries, September 1972; to BSC Steelworks Disposal Site, Corby, December 1980; to North Yorkshire Moors Railway, Grosmont, 16th March 1981; to Great Central Railway, Loughborough, 11th December 1984.</td></tr>
<tr><td>D9530</td><td>Swindon</td><td></td><td>1965</td><td>86A</td><td>10/68</td><td>F</td><td>NFT</td></tr>
<tr><td colspan="8">to Gulf Oil Co Ltd, Waterston, Pembrokeshire, 26th September 1969; to NCB Mardy Colliery, Glamorgan, via BR Canton Depot, 28th October 1975; to BR Canton Depot and BR Ebbw Junction Depot, for repairs, July 1976; to Mardy Colliery, August 1976; to BR Canton Depot, 2nd October 1977; to Mardy Colliery, about December 1977; scrapped March 1982.</td></tr>
<tr><td>D9531</td><td>Swindon</td><td></td><td>1965</td><td>86A</td><td>12/67</td><td>P</td><td>No.31/2100-523</td></tr>
<tr><td colspan="8">to Arnott Young Ltd, Parkgate, Rotherham, July 1968; despatched from 85A; to Hargreaves Industrial Services Ltd, NCBOE British Oak Disposal Point, Crigglestone, about November 1968; to NCB Burradon Colliery, October 1973; to Ashington Colliery, by April 1974; to Lambton Engine Works, Philadelphia, about September 1981; to Ashington Colliery, 27th October 1981.</td></tr>
<tr><td>D9532</td><td>Swindon</td><td></td><td>1965</td><td>50B</td><td>4/68</td><td>F</td><td>57/8311-37</td></tr>
<tr><td colspan="8">to Stewarts & Lloyds Minerals Ltd, Corby Quarries, November 1968; to BSC Steelworks Disposal Site, Corby, December 1980; scrapped on site by Shanks & McEwan Ltd of Corby, February 1982.</td></tr>
<tr><td>D9533</td><td>Swindon</td><td></td><td>1965</td><td>50B</td><td>4/68</td><td>F</td><td>47/8311-26</td></tr>
<tr><td colspan="8">to Stewarts & Lloyds Minerals Ltd, Corby Quarries, December 1968; to BSC Steelworks Disposal Site, Corby, December 1980; scrapped on site by Shanks & McEwan Ltd of Corby, September 1982.</td></tr>
<tr><td>D9534</td><td>Swindon</td><td></td><td>1965</td><td>50B</td><td>4/68</td><td>F</td><td>ECCLES</td></tr>
<tr><td colspan="8">to APCM, Hope, Derbyshire, October 1968; sold for export (see Appendix C); left Hope 5th May 1975.</td></tr>
<tr><td>D9535</td><td>Swindon</td><td></td><td>1965</td><td>86A</td><td>12/68</td><td>F</td><td>37/9312-59</td></tr>
<tr><td colspan="8">to NCB Burradon Colliery, November 1970; to Ashington Central Workshops, about July 1973; to Burradon Colliery, about March 1974; to Weetslade Loco Shed, January 1976; to Backworth Colliery, about May 1976; to Ashington Colliery, September 1980; to Lambton Engine Works, Philadelphia, December 1980; to Ashington Colliery, 23rd April 1981; withdrawn November 1983; scrapped on site by D. Short Ltd of North Shields, January 1984.</td></tr>
<tr><td>D9536</td><td>Swindon</td><td></td><td>1965</td><td>87A</td><td>4/69</td><td>F</td><td>No.5/9312-91</td></tr>
<tr><td colspan="8">to NCB Ashington Colliery, March 1970; to BR Gosforth Depot for wheel turning, August 1973; returned to Ashington Colliery; to Lambton Engine Works, Philadelphia, January 1977; to Ashington Colliery, May 1977; to Lambton Engine Works, January 1978; to Ashington Colliery, June 1978; scrapped on site, week ending 30th November 1985.</td></tr>
</table>

<table>
<tr><th>BR number</th><th>Builder</th><th>Works number</th><th>Year built</th><th>Last BR Shed</th><th>Date wdn</th><th>P/F</th><th>Title</th></tr>
<tr><td>D9537</td><td>Swindon</td><td></td><td>1965</td><td>50B</td><td>4/68</td><td>P</td><td>D9537</td></tr>
<tr><td colspan="8">to Stewarts & Lloyds Minerals Ltd, Corby Quarries, November 1968; to BSC Penn Green Crane Depot for storage, about November 1981; to Gloucestershire & Warwickshire Railway Society, Toddington, 23rd November 1982.</td></tr>
<tr><td>D9538</td><td>Swindon</td><td></td><td>1965</td><td>87A</td><td>4/69</td><td>F</td><td>160</td></tr>
<tr><td colspan="8">to Shell-Mex & BP Ltd, Shell Haven, Essex, April 1970; returned to BR, Swindon; resold to BSC; to Ebbw Vale Steelworks, Monmouthshire, February 1971; to BSC Corby Quarries, April 1976; scrapped on site by Shanks & McEwan Ltd of Corby, September 1982.</td></tr>
<tr><td>D9539</td><td>Swindon</td><td></td><td>1965</td><td>50B</td><td>4/68</td><td>P</td><td>51/8311-30</td></tr>
<tr><td colspan="8">to Stewarts & Lloyds Minerals Ltd, Corby Quarries, October 1968; to BSC Steelworks Disposal Site, Corby, December 1980; to Gloucestershire & Warwickshire Railway, Toddington, 23rd February 1983.</td></tr>
<tr><td>D9540</td><td>Swindon</td><td></td><td>1965</td><td>50B</td><td>4/68</td><td>F</td><td>36/508/2233-508</td></tr>
<tr><td colspan="8">to NCB Philadelphia Loco Shed, County Durham, November 1968; to Burradon Colliery, November 1971; to Ashington Colliery, June 1972; to Burradon Colliery, by April 1974; to Weetslade Loco Shed, January 1976; withdrawn November 1983; scrapped on site by D. Short Ltd of North Shields, 11th January 1984.</td></tr>
<tr><td>D9541</td><td>Swindon</td><td></td><td>1965</td><td>50B</td><td>4/68</td><td>F</td><td>66</td></tr>
<tr><td colspan="8">to Stewarts & Lloyds Minerals Ltd, Harlaxton Quarries, Lincolnshire, November 1968; to Corby Quarries, August 1974; to BSC Steelworks Disposal Site, Corby, December 1980; scrapped on site by Shanks & McEwan Ltd of Corby. August 1982.</td></tr>
<tr><td>D9542</td><td>Swindon</td><td></td><td>1965</td><td>50B</td><td>4/68</td><td>F</td><td>48/8311-27</td></tr>
<tr><td colspan="8">to Stewarts & Lloyds Minerals Ltd, Corby Quarries, December 1968; to BSC Steelworks Dispoal Site, Corby, December 1980; scrapped on site by Shanks & McEwan Ltd of Corby, August 1982.</td></tr>
<tr><td>D9544</td><td>Swindon</td><td></td><td>1965</td><td>50B</td><td>4/68</td><td>F</td><td>D9544</td></tr>
<tr><td colspan="8">to Stewarts & Lloyds Minerals Ltd, Corby Quarries, 2nd November 1968; dismantled in 1970 and used for spares; remains scrapped in September 1980.</td></tr>
<tr><td>D9545</td><td>Swindon</td><td></td><td>1965</td><td>50B</td><td>4/68</td><td>F</td><td>D9545</td></tr>
<tr><td colspan="8">to NCB Ashington Colliery, November 1968; later dismantled for spares; remains scrapped in July 1979.</td></tr>
<tr><td>D9547</td><td>Swindon</td><td></td><td>1965</td><td>50B</td><td>4/68</td><td>F</td><td>49/8311-28</td></tr>
<tr><td colspan="8">to Stewarts & Lloyds Minerals Ltd, Corby Quarries, December 1968; to BSC Steelworks Disposal Site, Corby, December 1980; scrapped on site by Shanks & McEwan Ltd of Corby, August 1982.</td></tr>
<tr><td>D9548</td><td>Swindon</td><td></td><td>1965</td><td>50B</td><td>4/68</td><td>F</td><td>67/8411-27</td></tr>
<tr><td colspan="8">to Stewarts & Lloyds Minerals Ltd, Harlaxton Quarries, Lincolnshire, November 1968; to Corby Quarries, August 1974; to BSC Steelworks Disposal Site, Corby, December 1980; to Hunslet Engine Co Ltd, Leeds, 19th November 1981; overhauled and rebuilt to 5ft 6in gauge; exported to Spain via Goole Docks, June 1982.</td></tr>
<tr><td>D9549</td><td>Swindon</td><td></td><td>1965</td><td>50B</td><td>4/68</td><td>F</td><td>64/8311-33</td></tr>
<tr><td colspan="8">to Stewarts & Lloyds Minerals Ltd, Corby Quarries, November 1968; to Glendon Quarry, October 1973; to Corby Quarries, 26th June 1974; to BSC Tube Works, Corby, September 1980; to Hunslet Engine Co Ltd, Leeds, 14th November 1981; overhauled and rebuilt to 5ft 6in gauge; exported to Spain via Goole Docks, June 1982.</td></tr>
<tr><td>D9551</td><td>Swindon</td><td></td><td>1965</td><td>50B</td><td>4/68</td><td>P</td><td>50/8311-29</td></tr>
<tr><td colspan="8">to Stewarts & Lloyds Minerals Ltd, Corby Quarries, December 1968; to BSC Tube Works, Corby, July 1980; to West Somerset Railway, Taunton, 5th June 1981.</td></tr>
<tr><td>D9552</td><td>Swindon</td><td></td><td>1965</td><td>50B</td><td>4/68</td><td>F</td><td>59/8411-21</td></tr>
<tr><td colspan="8">to Stewarts & Lloyds Minerals Ltd, Buckminster Quarries, Lincolnshire, September 1968; to Corby Quarries, June 1972; scrapped September 1980.</td></tr>
</table>

BR number	*Builder*	*Works number*	*Year built*	*Last BR Shed*	*Date wdn*	*P/F*	*Title*
D9553	Swindon		1965	50B	4/68	P	54/8311-34
to Stewarts & Lloyds Minerals Ltd, Corby Quarries, November 1968; to BSC Steelworks Disposal Site, Corby, December 1980; to Gloucestershire & Warwickshire Railway Society, Toddington, 23rd February 1983.							
D9554	Swindon		1965	50B	4/68	F	58/8311-38
to Stewarts & Lloyds Minerals Ltd, Corby Quarries, November 1968; to BSC Steelworks Disposal Site, Corby, December 1980; scrapped on site by Shanks & McEwan Ltd of Corby, August 1982.							
D9555	Swindon		1965	87A	4/69	P	D9555/9107-57
to NCB Burradon Colliery, March 1970; despatched from 86A; to Ashington Colliery, February 1975; to Backworth Colliery, December 1975; to Ashington Colliery, August 1980.							

SECTION 19

BR number	*Builder*	*Works number*	*Year built*	*Last BR Shed*	*Date wdn*	*P/F*	*Title*
12049	Derby		1948	1E	10/71	P	NPT
to Day & Sons (Brentford) Ltd, Brentford Town Goods Depot, London, October 1972; to BR Old Oak Common Depot for repairs, from November 1976 to February 1977; returned to Day & Sons.							
12050	Derby		1949	9A	7/70	F	12050
to NCB Philadelphia Loco Shed, County Durham, April 1971; dismantled for spares, June 1971; remains scrapped in June 1972.							
12052	Derby		1949	5A	6/71	P	MP228
to Derek Crouch (Contractors) Ltd, NCBOE Widdrington Disposal Point, Northumberland, December 1971.							
12054	Derby		1949	6A	7/70	F	12054
to A.R. Adams Ltd, Newport, September 1971; used as a hire loco (see Appendix A); scrapped April 1984.							
12060	Derby		1949	9A	2/71	F	512/2233-512
to NCB Derwenthaugh Loco Shed, Blaydon, County Durham, March 1971; to Philadelphia Loco Shed, April 1971; scrapped on site by C.F. Booth Ltd of Rotherham, November 1985.							
12061	Derby		1949	8J	10/71	P	4
to NCB Nantgarw Coking Plant, Glamorgan, 11th December 1972; despatched from 8F; to BR Canton Depot for repairs, December 1974; to Nantgarw, December 1974; to BR Swindon Works for repairs, 10th December 1981; to Nantgarw, 8th February 1982.							
12063	Derby		1949	8F	1/72	P	5
to NCB Nantgarw Coking Plant, Glamorgan, 11th December 1972; to BR Canton Depot for repairs, January 1977; returned to Nantgarw.							
12071	Derby		1950	8F	10/71	P	6
to NCB Nantgarw Coking Plant, Glamorgan, 11th December 1972; to BR Canton Depot for repairs, about December 1974; to Nantgarw, December 1974; to Canton Depot, about November 1976; to Nantgarw, November 1976; to BR Ebbw Junction Depot, March 1977; to BR Swindon Works, 6th July 1977; to Nantgarw, October 1977; to BR Swindon Works, September 1980; to Nantgarw, January 1981.							
12074	Derby		1950	6A	1/72	P	12074
to Johnsons (Chopwell) Ltd, NCBOE Swalwell Disposal Point, County Durham, June 1972; despatched from 5A.							
12077	Derby		1950	8F	10/71	P	12077
to Cashmore Ltd, Great Bridge, Staffordshire, September 1973; to Midland Railway Company, Butterley, Derbyshire, 16th December 1978.							
12082	Derby		1950	6G	10/71	P	12082
to Shellstar (UK) Ltd, Ince Marshes, Ellesmere Port, March 1973; despatched from 6A; loaned to Manchester Ship Canal Co, Ellesmere Port from 16th July 1974 to 25th October 1974; to BR Swindon Works for repairs, February 1978; to Shellstar, March 1978.							

<table>
<tr><th>BR number</th><th>Builder</th><th>Works number</th><th>Year built</th><th>Last BR Shed</th><th>Date wdn</th><th>P/F</th><th>Title</th></tr>
<tr><td>12083</td><td>Derby</td><td></td><td>1950</td><td>12A</td><td>10/71</td><td>P</td><td>12083</td></tr>
<tr><td colspan="8">to Tilcon Ltd, Swinden Works, Grassington, July 1973; to BR Doncaster Depot for repairs, September 1974; to Tilcon, October 1974.</td></tr>
<tr><td>12084</td><td>Derby</td><td></td><td>1950</td><td>5A</td><td>5/71</td><td>F</td><td>514/2233-514</td></tr>
<tr><td colspan="8">to NCB Burradon Colliery, October 1971; to Philadelphia Loco Shed, County Durham, November 1971; to Silksworth Colliery, April 1972; to Hylton Colliery, June 1972; to Philadelphia Loco Shed, March 1975; to Easington Colliery, December 1975; to Blackhall Colliery, January 1976; to Bates Colliery, Blyth, April 1976; to Lambton Engine Works, Philadelphia, 25th February 1982; to Philadelphia Loco Shed, 21st October 1983; scrapped on site by C.F. Booth Ltd of Rotherham, November 1985.</td></tr>
<tr><td>12085</td><td>Derby</td><td></td><td>1950</td><td>12A</td><td>5/71</td><td>F</td><td>12085</td></tr>
<tr><td colspan="8">to Thos. W. Ward Ltd, Barrow-in-Furness, May 1973; scrapped about June 1976.</td></tr>
<tr><td>12088</td><td>Derby</td><td></td><td>1951</td><td>8J</td><td>5/71</td><td>P</td><td>12088</td></tr>
<tr><td colspan="8">to Johnsons (Chopwell) Ltd, NCBOE Swallwell Disposal Point, County Durham, July 1972; despatched from 8F.</td></tr>
<tr><td>12093</td><td>Derby</td><td></td><td>1951</td><td>5A</td><td>5/71</td><td>P</td><td>MP229</td></tr>
<tr><td colspan="8">to Derek Crouch (Contractors) Ltd, NCBOE Widdrington Disposal Point, Northumberland, December 1971.</td></tr>
<tr><td>12098</td><td>Derby</td><td></td><td>1952</td><td>8F</td><td>2/71</td><td>P</td><td>513/2233-513</td></tr>
<tr><td colspan="8">to NCB Derwenthaugh Loco Shed, County Durham, March 1971; to Philadelphia Loco Shed, April 1971; to National Smokeless Fuels Ltd, Lambton Coking Plant, about July 1985; to Tyne & Wear Museum, 5th January 1987.</td></tr>
<tr><td>12099</td><td>Derby</td><td></td><td>1952</td><td>1E</td><td>7/71</td><td>P</td><td>NPT</td></tr>
<tr><td colspan="8">to Murphy Bros Ltd, NCBOE Lion Disposal Point, Blaenavon, about April 1972; to Taylor Woodrow Construction Ltd, NCBOE Cwm Bargoed Disposal Point, October 1975; to Hargreaves Industrial Services Ltd, NCBOE British Oak Disposal Point, Crigglestone, August 1981; to NCBOE Bowers Row Disposal Point, about February 1983.</td></tr>
<tr><td>12119</td><td>Darlington</td><td></td><td>1952</td><td>50B</td><td>11/68</td><td>F</td><td>509/2233-509</td></tr>
<tr><td colspan="8">to NCB Philadelphia Loco Shed, County Durham, February 1969; to Lambton Engine Works, Philadelphia, November 1980; to Philadelphia Loco Shed, January 1981; scrapped on site by C.F. Booth Ltd of Rotherham, November 1985.</td></tr>
<tr><td>12120</td><td>Darlington</td><td></td><td>1952</td><td>50B</td><td>12/68</td><td>F</td><td>510</td></tr>
<tr><td colspan="8">to NCB Philadelphia Loco Shed, County Durham, February 1969; to Whittle Colliery, June 1978; to Lambton Engine Works, Philadelphia, August 1979; scrapped March 1980.</td></tr>
<tr><td>12122</td><td>Darlington</td><td></td><td>1952</td><td>40B</td><td>7/71</td><td>F</td><td>12122</td></tr>
<tr><td colspan="8">to Murphy Bros Ltd, NCBOE Lion Disposal Point, Blaenavon, January 1972; to Taylor Woodrow Construction Ltd, NCBOE Cwm Bargoed Disposal Point, October 1975; to Hargreaves Industrial Services Ltd, NCBOE British Oak Disposal Point, Crigglestone, for spares only, August 1981; scrapped on site by Rawden's of Barnsley, October 1985.</td></tr>
<tr><td>12131</td><td>Darlington</td><td></td><td>1952</td><td>30A</td><td>3/69</td><td>P</td><td>12131</td></tr>
<tr><td colspan="8">to NCB Betteshanger Colliery, Kent, March 1969; to Snowdown Colliery, June 1976; to North Norfolk Railway, 25th April 1982.</td></tr>
<tr><td>12133</td><td>Darlington</td><td></td><td>1952</td><td>40B</td><td>1/69</td><td>F</td><td>511/2100-526</td></tr>
<tr><td colspan="8">to NCB Philadelphia Loco Shed, County Durham, May 1969; to Lambton Engine Works, Philadelphia, 1979; to Whittle Colliery, about March 1981; to Lambton Engine Works, 23rd April 1981; to Philadelphia Loco Shed, 13th August 1981; scrapped on site by C.F. Booth Ltd of Rotherham, November 1985.</td></tr>
</table>

BR number	*Builder*	*Works number*	*Year built*	*Last BR Shed*	*Date wdn*	*P/F*	*Title*

SECTION 20

15222	Ashford		1949	73C	10/71	F	15222

to Cashmore Ltd, Newport, May 1972; to John Williams Ltd, Blaenyfan, Kidwelly, 1974, where used as a stationary generator; there in March 1978, but scrapped later in 1978.

15224	Ashford		1949	75C	10/71	P	15224/1802-B5

to NCB Betteshanger Colliery, Kent, October 1972; despatched from 75A; to Snowdown Colliery, 27th May 1976; left Snowdown on 9th October 1982 and stored in BR Hove Goods Yard; to Brighton Works Locomotive Association, Preston Park Car Sheds, Brighton, April 1983; to D. Milham, Lavender Line, June 1985.

15231	Ashford		1951	73F	10/71	F	TILCON

to Tilcon Ltd, Swinden Works, Grassington, June 1972; scrapped January 1984.

SECTION 21

ED10	RH	411322	1958	BSD	2/65	P	E9

to Thos. W. Ward Ltd, September 1965; to Cleveland Bridge & Engineering Co Ltd, Darlington, May 1966; used on the Tinsley Viaduct, Sheffield, contract; to Shepherd Hill & Co Ltd, November 1969; to Trackbed Hovercraft Ltd, Earith, Huntingdonshire, about May 1973; here converted to operate on Hovertrack and fitted with rubber tyres; to E. Hampton, St Ives, Huntingdonshire, 1975.

SECTION 22

ZM32	RH	416214	1957	HW	3/64	P	NPT

to S.E.E.C. Manchester, September 1965; sold to a buyer in British Honduras, but sale cancelled and the loco stored at Liverpool Docks; resold for preservation to R.P. Morris, Longfield, Kent, December 1971; rebuilt to 2ft 0in gauge by Alan Keef Ltd, Cote, Oxfordshire; moved to The Narrow Gauge Railway Centre of North Wales, Gloddfa Ganol, Blaenau Ffestiniog, Gwynedd, May 1978.

SECTION 23

D0226	EE VF	2345 D226	1956		10/60	P	D0226/VULCAN

Given trials on BR 1956–1957; returned to English Electric Ltd, Vulcan Works, Newton-le-Willows, 1957; stored; to Keighley & Worth Valley Railway, Haworth, March 1966; to BR Doncaster Depot for wheel turning, 4th March 1979; returned to KWVR.

SECTION 24

85049	RH	393325	1956	CJ	c4/86	P	85049

Used by BR at Chesterton Junction PW Depot, Cambridge; to Northamptonshire Ironstone Railway Trust, 2nd August 1986.

85051	RH	404967	1957	CJ	c4/86	P	85051

Used by BR at Chesterton Junction PW Depot, Cambridge; to Cadeby Rectory, Market Bosworth, Leicestershire, 3rd July 1986.

APPENDIX A

A.R. Adams Ltd, Robert Street, Newport, Gwent

This company hired locomotives to various concerns, mainly collieries and coal depots in South Wales. Between hirings the locos were stored at the premises of Rowecord Engineering Ltd, West Side, Old Town Dock, Newport, Gwent. Details of known hirings are given below:—

D2139 to NCB Marketing Department, Gwent Coal Concentration Depot, Newport, from about March 1969 to July 1969; to NCB Coal Products Division, Nantgarw, from August 1970 to about December 1970.

D2178 to NCB Aberaman Colliery, January 1970; to Wiggins Teape Ltd, Ely Paper Works, Cardiff, from February 1970, to 1971; to Powell Duffryn Fuels Ltd, NCBOE Gwaun-cae-Gurwen Disposal Point, from about March 1972 to July 1972.

D2181 to NCB Marketing Department, Gwent Coal Concentration Depot, Newport, from about March 1969, and to whom the loco was sold by August 1971.

D2182 to NCB Coal Products Division, Caerphilly Tar Works, from January 1969 to February 1969; to Sir Lindsay Parkinson & Co Ltd, Glyn Neath, February 1969, and to whom the loco was subsequently sold.

D2186 to NCB Aberaman Colliery, from February 1970 to about October 1970; to NCB Tower Colliery from March 1971 to about September 1972; scrapped January 1981.

D2193 to Powell Duffryn Fuels Ltd, NCBOE Coed Bach Disposal Point, Kidwelly, from September 1969 to about June 1970; to NCB Coal Products Division, Coed Ely, Tonyrefail, from July 1970 to about September 1970; to NCB Mountain Ash Colliery from about February 1971 to about April 1971; to Powell Duffryn Fuels Ltd, NCBOE Coed Bach Disposal Point, Kidwelly, during 1972; to NCB Coal Products Division, Nantgarw, from 1972 to 1973; to NCB Taff Merthyr Colliery, from August 1973 to December 1973; to NCB Garw Colliery, from October 1977 to September 1978; scrapped January 1981.

D2244 to Monsanto Chemicals Ltd, Newport, Monmouthshire, from August 1970 to about February 1971; to NCB Coed Cae, Pencoed, from April 1971 to March 1972; to NCB Ogmore Central Washery, from March 1972 to May 1972; scrapped January 1981.

12054 to NCB Mountain Ash Colliery, from about September 1971 to May 1972; to NCB Tower Colliery, Hirwaun, from May 1972 to October 1973; to NCB Mardy Colliery, from April 1974 to about October 1975; to NCB Mardy Colliery, from about March 1976 to September 1979; to BR Canton Depot for repairs, September 1979; to NCB Mardy Colliery from about November 1979 to March 1981; scrapped April 1984.

APPENDIX B

T.J. Thomson & Son Ltd, Stockton, Cleveland

This company purchased three ex-BR Departmental locomotives, which were taken to their Millfield Scrap Works at Stockton. These were never used at this site, but were stored for several years awaiting scrapping. All three were despatched to Thomson's yard from BR Thornaby Depot, in May 1970, having been stored at that location from about August 1969. The locomotives were all 4-wheel diesels of the maker's class 88DS, and were fitted with 88bhp engines, and mechanical transmission. They were scrapped in October 1981.

Departmental number	*Builder*	*Works number*	*Year built*	*Last working location*
56	RH	338424	1955	Etherley Tip
82	RH	425485	1959	Dinsdale Welded Rail Depot
87	RH	463152	1961	Geneva Yard, Darlington

APPENDIX C

Locomotives sold abroad

BR number	*Builder*	*Works number*	*Year built*	*Last BR Shed*	*Date wdn*	*P/F*	*Title*
D2010	Swindon		1958	51L	11/74	P	Not known
03010	to Shipbreakers (Queenborough) Ltd, Kent; exported from Stranraer, May 1976; to Trieste, Italy.						
D2019	Swindon		1958	32A	7/71	P	Not known
	to Shipbreakers (Queenborough) Ltd, Kent; exported September 1972; to Stabilimento ISA, Ospitaletto, Brescia, Italy.						
D2032	Swindon		1958	32A	7/71	P	Not known
	to Shipbreakers (Queenborough) Ltd, Kent; exported August 1972; to Stabilimento ISA, Ospitaletto, Brescia, Italy.						
D2033	Swindon		1958	32A	12/71	P	Not known
	to Shipbreakers (Queenborough) Ltd, Kent; exported August 1972; to Siderveica, Montirone, Brescia, Italy.						
D2036	Swindon		1959	32A	12/71	P	Not known
	to Shipbreakers (Queenborough) Ltd, Kent; exported August 1972; to Stabilimento ISA, Ospitaletto, Brescia, Italy.						
D2081	Doncaster		1960	31B	12/80	P	Not known
03081	Sold from BREL Swindon Works; exported to Belgium, November 1981.						
D2098	Doncaster		1960	51A	11/75	P	Not known
03098	to Shipbreakers (Queenborough) Ltd, Kent; exported from Stranraer, May 1976; to Trieste, Italy.						
D2128	Swindon		1960	82A	7/76	F	Not known
03128	to Birds, Long Marston, October 1976; resold for export, probably as scrap or for spares; exported, minus its engine, from Harwich, December 1976.						
D2134	Swindon		1960	82A	7/76	P	Not known
03134	to Birds, Long Marston, October 1976; resold for export about January 1977; destination not known.						
D2153	Swindon		1960	51L	11/75	P	Not known
03153	to Shipbreakers (Queenborough) Ltd, Kent; exported from Stranraer, May 1976; to Trieste, Italy.						
D2156	Swindon		1960	52A	11/75	P	Not known
03156	to Shipbreakers (Queenborough) Ltd, Kent; exported from Stranraer, May 1976; to Trieste, Italy.						
D2157	Swindon		1960	50C	12/75	P	Not known
03157	to Shipbreakers (Queenborough) Ltd, Kent; exported February 1977; to Trieste, Italy.						
D2164	Swindon		1960	30A	1/76	P	Not known
03164	to Shipbreakers (Queenborough) Ltd, Kent; exported February 1977; to Trieste, Italy.						

BR number	Builder	Works number	Year built	Last BR Shed	Date wdn	P/F	Title
D2216	VF	D265	1955	30A	5/71	P	Not known
	DC	2539					
to Shipbreakers (Queenborough) Ltd, Kent; exported September 1972; to Stabilimento ISA, Ospitaletto, Brescia, Italy.							
D2232	VF	D282	1956	52A	3/68	P	Not known
	DC	2556					
to Shipbreakers (Queenborough) Ltd, Kent; exported August 1972; to Attilio Rossi, Rome, Italy.							
D2289	RSH	8122	1960	70D	9/71	P	Not known
	DC	2669					
to Shipbreakers (Queenborough) Ltd, Kent; exported April 1972; to Feralpi, Lanato, Brescia, Italy.							
D2295	RSH	8128	1960	70D	4/71	P	Not known
	DC	2675					
to Shipbreakers (Queenborough) Ltd, Kent; exported September 1972; to Acciaierie di Lonato, Lonato, Brescia, Italy.							
D2432	AB	459	1960	65A	12/68	P	Not known
to Shipbreakers (Queenborough) Ltd, Kent; exported March 1977; to Trieste, Italy.							
D2993	RH	480694	1962	70D	10/76	P	Not known
07009	to Shipbreakers (Queenborough) Ltd, Kent; exported March 1977; to Italy.						
D3047	Derby		1954	70D	7/73	P	Not known
Overhauled at BREL Derby; exported February 1975; to Lamco Mining Co, Liberia.							
D3092	Derby		1954	73C	10/72	P	Not known
Overhauled at BREL Derby; exported May 1974; to Lamco Mining Co, Liberia.							
D3094	Derby		1954	73F	10/72	P	Not known
Overhauled at BREL Derby; exported May 1974; to Lamco Mining Co, Liberia.							
D3098	Derby		1955	73F	10/72	P	Not known
Overhauled at BREL Derby; exported May 1974; to Lamco Mining Co, Liberia.							
D3100	Derby		1955	75C	10/72	P	Not known
Overhauled at BREL Derby; exported May 1974; to Lamco Mining Co, Liberia.							
D3639	Darlington		1958	36A	7/69	P	Not known
to Conakry, Guinea, West Africa; exported March 1970.							
D3649	Darlington		1959	36A	7/69	P	Not known
to Conakry, Guinea, West Africa; exported March 1970.							
D9505	Swindon		1964	50B	4/68	P	Not known
to Sobemai, Maldegem, near Bruges, Belgium; exported from Harwich, May 1975; believed later sold to an industrial user in Belgium.							
D9515	Swindon		1964	50B	4/68	P	Not known
to Spain; exported from Goole Docks, June 1982.							
D9534	Swindon		1965	50B	4/68	P	Not known
to Sobemai, Maldegem, near Bruges, Belgium; exported from Harwich, May 1975; sold to an industrial user near Milan, Italy, 1976.							
D9548	Swindon		1965	50B	4/68	P	Not known
to Spain; exported from Goole Docks, June 1982.							
D9549	Swindon		1965	50B	4/68	P	Not known
to Spain; exported from Goole Docks, June 1982.							